TARTE
by GARUHARU

GARUHARU MASTER BOOK SERIES 2

TARTE by GARUHARU

초판 1쇄 발행	2020년 7월 1일
초판 6쇄 발행	2023년 10월 10일

지은이	윤은영
어시스턴트	김은솔
영문번역	김예성
펴낸이	한준희
발행처	(주)아이콕스

기획·편집	박윤선
디자인	김보라
사진	박성영
스타일링	이화영
영업·마케팅	김남권, 조용훈, 문성빈
경영지원	김효선, 이정민

주소	경기도 부천시 조마루로385번길 122 삼보테크노타워 2002호
홈페이지	www.icoxpublish.com
쇼핑몰	www.baek2.kr (백두도서쇼핑몰)
인스타그램	@thetable_book
이메일	thetable_book@naver.com
전화	032) 674-5685
팩스	032) 676-5685
등록	2015년 07월 09일 제 386-251002015000034호
ISBN	979-11-6426-125-3

TARTE
by GARUHARU

타 르 트 바 이 가 루 하 루

윤은영 지음

더 테이블
THE TABLE

PROLOGUE

열다섯. 작은 손으로 빚어낸 밀가루 반죽이 오븐 속에서 근사한 과자로 부풀어 오르는 모습을 지켜보았던 날, 제 마음도 따뜻하고 달콤한 과자처럼 부풀어 올랐습니다. 울퉁불퉁 오트밀 쿠키 한 조각을 맛보고 엄지를 추켜세우며 한껏 웃던 가족들과 친구들의 행복한 표정이 저를 '파티시에'라는 길로 이끌어 주었습니다.

파티시에는 다른 이들에게 달콤한 맛의 휴식과 즐거움을 주는 행복한 직업이지만, 기술을 배우고 능숙해지기까지 긴 시간과 노력이 필요했습니다. 하지만 이 길고 힘든 시간 끝에는 저와 제가 만든 제품들이 함께 성장해 있었습니다.

단지 레시피가 아닌 그 과정에서 겪었던 많은 시행착오와 실패를 경험하며 터득한 포인트와 팁, 도구 활용법은 물론 제조 공정에서의 잦은 실수를 줄여주는 방법들을 이 책에 담았습니다. 또한 해외 마스터 클래스를 진행하며 만난 다양한 문화권의 훌륭한 셰프들과 새로운 식재료들로부터 받은 영감의 결과물을 오롯이 담고자 노력했습니다.

이 책이 누군가에게 새로운 영감을 주는 도구로 사용되길 바랍니다. 이 책의 레시피를 토대로 여러분의 주변에서 구할 수 있는 재료를 활용해 다양한 시도를 해보셨으면 좋겠습니다. 제가 늘 작업대 위에서 기존의 제품과 새로운 재료의 조합을 고민하듯, 여러분의 작업실에서도 이러한 고민과 새로운 시도가 계속되길 희망하며, 그 과정에서 저의 책이 작은 도움이 되었으면 합니다.

끝으로 이 책을 펴내는 데 도움을 주신 많은 분들과 아낌없는 지원을 해주신 더테이블 관계자 분들께 고마운 마음을 전합니다.

윤은영

Fifteen. The day I watched the dough I made with my small hands rise up into gorgeous cookies, my heart billowed just like the warm and sweet cookie. The happy faces of family and friends who tasted a piece of the lumpy oatmeal cookie, raising their thumbs up with big loving laughers, led me to the path of becoming a 'pâtissier.'

Being a pâtissier a blissful job, giving others a sweet taste of relaxation and delightfulness, but it took a long time and effort to learn the skills and become proficient. However, at the end of this long and challenging time, I and the products I made have been growing together.

This book is not just a collection of recipes, but also reflects a lot of trials and errors that we have experienced along the way, and tried to cover the points and tips we've learned, including how to utilize tools, as well as ways to minimize frequent mistakes during the process of making. I also tried to capture the results of inspirations from the talented chefs we've met from various cultures and new ingredients we came across while conducting the overseas master classes.

I hope this book will serve as a new inspiration for someone. Please make a variety of attempts with ingredients available near you by utilizing the recipes in this book. Just as I always contemplate the combination of existing products and new ingredients on my workstation, I hope these thoughts and further attempts will continue in your studio and wish my book will be of little help along the way.

And last but not least, I would like to thank many people who helped me publish this book and the officials of THETABLE publishers for their generous support.

Yun Eunyoung

HOW TO USE THIS BOOK
이 책을 활용하는 법

타르트는 다양한 종류의 충전물을 사용하는 제과이므로 이 책에서는 충전물의 종류에 따라 컨벡션 오븐, 데크 오븐으로 나누어 사용하였습니다.
컨벡션 오븐과 데크 오븐에 대한 설명은 본책 37p를 참고하시기 바랍니다.

Tarte is a confectionery that uses various types of fillings, so in this book, a convection oven and deck oven are used accordingly depending on the type of fillings. Please refer to page 37 of this book for explanations on the convection oven and deck oven.

이 책에서는 사블레 반죽, 브리제 반죽, 글루텐 프리 반죽 등 타르트에 활용할 수 있는 6가지 반죽을 제시합니다. 레시피에 따라 취향에 따라 선택할 수 있습니다.

Six types of doughs are presented, which can be used for tartes; sablé dough, brisée dough, gluten-free dough, etc. It can be chosen according to preference or recipe.

'지름 7cm 타르트 틀 기준 15개의 타르트 셸이 완성된다'는 의미입니다.

It means 'this recipe makes 15 pieces of tarte shells with 7cm diameter tarte mold.'

¹ PATE SABLER
파트 아 사블레

Ø 7cm
15ea

CONVECTION OVEN
160°C

ingredients

드라이버터 86g	86g Dry butter
중력분 163g	163g All-purpose flour
슈거파우더 82g	82g Powdered sugar
소금 0.9g	0.9g Salt
달걀전란 48g	48g Whole eggs
아몬드파우더 12g	12g Almond powder
세몰리나-서(중간 굵기) 12g	12g Semolina (medium grind)
전분 47g	47g Starch

이 책에서는 밀가루를 사용하는 일반적인 사블레 반죽, 브리제 반죽에 더해 쌀가루를 사용하는 글루텐 프리 스위트 반죽(53p)과 글루텐 프리 파이 반죽(61p)까지 총 6가지 타르트 반죽 레시피를 소개합니다. 기호에 따라 원하는 반죽을 선택할 수 있습니다.
또한 넛트 프리 아이콘이 표시되어 있는 레시피의 경우 견과류 알러지가 있으신 분들도 즐기실 수 있습니다.

In this book, a total of 6 types of tarte dough recipes are introduced; general sablé doughs and brisée dough using wheat flour, gluten-free sweet dough (53p), and gluten free pie dough (61p) using rice flour. You can use any dough of your choice. Additionally, the recipes with 'nut-free' icon can be enjoyed by those who are allergic to nuts.

지름 16cm 타르트 셸 기준 2개의 타르트가 완성되는 배합입니다. 필요한 만큼 배합을 줄이거나 늘려 사용할 수 있습니다.

This recipe makes two tartes with 16cm diameter tarte shells. The recipe can be divided or multiplied as needed.

¹⁹ SEAFOOD & TOMATO QUICHE

해산물 & 토마토 키슈

S - Ø 16cm 2ea

ingredients

파트 아 브리제(57p) or
파이 반죽 - 글루텐 프리
(61p)

GLUTEN FREE   NUTS FREE

아파레유			가르니튀르			장식물	
		25g	관자		크림치즈	바질	
			새우		체다치즈	드라이토마토	
			양파		고다치즈		
			마늘		올리브오일		
			감자		소금		
			방울토마토		후추		
			브로콜리		로즈마리		
			청양고추				

GARNITURE		DECORATION	
Scallops	Cream cheese	Basil	
Shrimps	Cheddar cheese	Dried tomatoes	
Onion	Gouda cheese		
Garlic	Olive oil		
Potatoes	Salt		
Cherry tomatoes	Ground pepper		
Broccoli	Rosemary		
Cheongyang pepper (or other spicy chili pepper)			

Fruits

Nut & Chocolate

Aroma

Cream & Milk

Savory

Preparation

재료 Ingredients

달걀 (EGGS)

우유 (MILK)

소금 (SALT)

생크림 (FRESH CREAM)

바닐라빈 (VANILLA BEAN)

버터 (BUTTER)

달걀

포장지에 적힌 산란일자와 포장일을 확인하여 신선한 달걀을 구입하는 것이 중요합니다. 달걀을 사용하기 전에는 물로 깨끗이 씻어 표면에 묻은 이물질을 제거한 후 사용하는 것을 권장합니다. 단, 물로 씻은 달걀을 보관하는 경우에는 세균이나 바이러스 균이 침투할 수 있으므로 필요한 만큼만 세척하여 사용하는 것이 좋습니다.

우유

우유는 제과에 있어 제품의 영양가를 높여줍니다. 또한 단백질과 유당을 함유해 메일라드 반응을 촉진시켜 과자 껍질의 색을 좋게 만듭니다. 고지방 우유, 저지방 우유, 칼슘 강화 우유처럼 특정 영양소를 강화한 우유, 유당을 소화하지 못하는 사람을 위한 락토프리 우유 등 다양한 제품이 있습니다. 이 책에서 사용한 우유는 모두 일반 우유입니다.

생크림

생크림은 식물성 기름이 첨가되지 않은 신선한 것을 구입해 유효기간 내에 사용하는 것이 좋습니다. 식물성 기름을 혼합한 가공 생크림은 작업성을 높여주고 보관 기간이 길지만 입안에서 잘 녹지 않고 풍미가 떨어지기 때문에 권장하지 않습니다. 생크림은 유지방 함량 20% 정도의 저지방 생크림부터 유지방 함량 45% 전후의 고지방 생크림까지 매우 다양합니다. 이책에서는 유지방 함량 38%의 동물성 생크림을 사용하였습니다.

소금

이 책에서 사용한 소금은 프랑스 서부, 대서양을 마주보는 게랑드 일대에서 생산되는 소금입니다. 염전 위에 꽃처럼 피어난 소금 결정을 하나하나 손으로 수확하여 '소금의 꽃(플뢰르 드 셀 fleur de sel)'이라 불립니다. 짠맛 뿐만 아니라 감칠맛까지 느낄 수 있으며 캐러멜과 같은 단맛과도 잘 어우러집니다.

바닐라빈

바닐라빈은 우유, 생크림, 버터와 같은 유제품 향의 조화를 높여주며 달걀 특유의 비릿함을 잡아줍니다. 긁어낸 바닐라빈은 크림이나 반죽에 넣어 향을 내고, 사용하고 남은 껍질은 설탕에 넣어 두었다가 설탕과 함께 푸드프로세서에 갈아 바닐라슈거를 만들어 사용할 수 있습니다.

버터

제과의 기본 재료인 버터는 냉동고나 냉장고에 넣어 차가운 상태, 실온에 꺼내두어 말랑한 상태, 전자레인지에 녹여 완전히 풀어진 상태 세 가지 형태로 사용합니다. 버터는 유지방 함량이 80% 이상인 퓨어버터를 사용하는 것이 좋습니다. 산화되기 쉬운 퓨어버터는 신선한 것을 구입해 빠른 시일 안에 사용하고, 오래 두고 사용해야 할 경우에는 냄새를 흡수하지 않도록 잘 밀폐하여 냉동 보관합니다. 이 책에서 사용한 버터는 제조 공정 중 유산균을 첨가해 숙성한 프랑스 천연 발효 버터입니다. 느끼함이 없고 유지방이 응축된 고소하고 담백한 맛을 느낄 수 있습니다.

Eggs

It is important to purchase fresh eggs by checking the date eggs were laid* and the packaging date printed on the carton. It is recommended to wash them thoroughly with water before using them to remove foreign substances on the surface. However, there is a possibility that bacteria or viruses may penetrate while storing eggs washed with water, so it's better to wash them only as needed.
* May not be indicated in some countries.

Milk

Milk improves the nutritional value of the baked goods in confectionery. It also contains protein and lactose, which promotes the Maillard reaction that helps to make an appetizing outer color of the confectionery goods. There are various kinds of milk, which include high-fat milk, low-fat milk, and milk with certain nutrients strengthened such as calcium-fortified milk, and lactofree milk for those who cannot digest lactose. The milk used in this book is regular whole milk.

Fresh cream (whipping cream)

It is recommended to purchase fresh cream that does not contain vegetable oils and use it within the expiration date. Processed whipping cream blended with vegetable oils increases workability and has a long shelf life; however, it does not melt nicely in the mouth and has poor flavor; therefore, it's not recommended. There are varieties of fresh creams ranging from low-fat fresh cream with a milk fat content of about 20% to high-fat fresh cream with a milk fat content of around 45%. In this book, we used fresh dairy cream containing 38% of milk fat.

Salt

The salt used in this book is produced in the region of Guerande, facing the Atlantic Ocean in western France. Each salt crystal blooming like flowers on the salt field are harvested by hand is called 'flower of salt (fleur de del).' You can feel not only its saltiness but also savory taste, and pairs well with sweet flavors such as caramels.

Vanilla beans

Vanilla bean enhances the harmony of dairy flavors such as milk, fresh cream, and butter and helps reduce the peculiar taste of eggs some may find fishy. The scraped vanilla bean seeds can be added to creams or batters to enhance flavor, and the remaining skin can be stored inside sugar, which can be ground all together with a food processor to make vanilla sugar.

Butter

Butter, which is the basis of confectionery ingredients, is used in three forms: cold state by keeping in a freezer or refrigerator, soft state by keeping in ambient temperature, and completely dissolved state by melting in a microwave. As for the butter, pure butter containing more than 80% of milk fat would be recommended to use. Pure butter should be purchased fresh and used as soon as possible because it's easy to oxidize. If it needs to be stored for a long time, it should be sealed tight, not to absorb any odor, and kept frozen. The butter used for this book is naturally fermented butter from France, aged by adding lactic acid bacteria during the manufacturing process. It doesn't taste greasy and gives a savory and clean flavor from concentrated milk fat.

마스카르포네 치즈
(MASCARPONE CHEESE)

펙틴 (PECTIN)

젤라틴매스 (GELATIN MASS)

럼 (RUM)

아가아가 (AGAR-AGAR)

젤라틴 (GELATIN)

마스카르포네 치즈

이탈리아의 대표적인 디저트 '티라미수'를 만드는 데 있어 필수적인 재료입니다. 이탈리아의 대표 치즈이며 지방 함량이 높아 섬세하고 부드러우면서도 진한 크림의 향을 느낄 수 있습니다.

럼

당밀과 사탕수수를 원료로 만든 증류주로 크림이나 과자의 풍미를 돋우는 데 도움을 줍니다.

펙틴

펙틴은 사과 찌꺼기나 감귤류의 껍질에서 추출한 천연 다당류입니다. 주로 크림이나 젤리의 질감을 완성하거나 안정화하는 데 사용되며 제조 과정 중에 음식물의 형체(바디감)와 질감(텍스처)을 만들어 주는 역할을 합니다. 펙틴은 일반적으로 하이 메톡실(HM), 로우 메톡실(LM) 두 가지 유형으로 나눌수 있으며 시중에는 다양한 종류의 펙틴이 존재합니다. 각각의 펙틴마다 물리화학적 특성이 다르므로 작업 전 공급자에게 제품 정보를 확인하는 것이 좋습니다.

젤라틴 & 젤라틴매스

응고제의 일종인 젤라틴은 무스나 젤리 등을 굳히는 데 사용되는 재료입니다. 소나 돼지의 뼈 또는 껍질에서 추출한 콜라겐을 정제해 건조시켜 만듭니다. 가루 타입과 얇은 판 형태의 제품이 있습니다. 이 책에서는 가루 형태의 젤라틴(200bloom)을 5배의 물과 혼합해 굳힌 후 사용하였습니다.(75p)

아가아가

한천을 주 원료로 하는 겔화제로 단단하면서도 탄력이 적은 겔을 형성할 수 있으며 겔화된 상태에서 다시 가열해 사용하는 것이 가능합니다.

Mascarpone cheese

This is an essential ingredient in making the classic Italian dessert called 'Tiramisu'. It is the representative cheese of Italy which has high fat content, giving a delicate and soft yet rich creamy aroma.

Rum

Rum is a distilled liquor made from molasses and sugar cane, which helps to enhance the flavor of cream or sweets.

Pectin

Pectin is a natural polysaccharide extracted from apple pomace or citrus peels. It is usaually used to complete or stabilize the texture of cream or jelly, and at the same time to complete the shape (body structure) and texture (sensory) of the product. Generally, there are two types of pectin; high methoxyl (HM) and low methoxyl (LM), and there are more types available in the market. Since each pectin has different physiochemical properties, it is recommended to check with the supplier for information about the product before use.

Gelatin & gelatin mass

Gelatine is a kind of coagulant that is used to harden mousse or jelly. It is made by refining and drying collagen extracted from bones or skins of cattle or pigs. There are powder type and thin sheet type of gelatin. For this book, powder type of gelatin (200 bloom) is mixed with 5 times of water and hardened to use (gelatin mass). (75p)

Agar-agar

A gelling agent, typically using algae as main ingredient, it is used to form a firm and less elastic gel, and can be reheated to use again.

타피오카 전분 (TAPIOCA STARCH)

슈거파우더 (SUGAR POWDER)

베이킹파우더
(BAKING POWDER)

드라이버터 (DRY BUTTER)

세몰리나 (SEMOLINA)

타피오카 전분

타피오카 전분은 글루텐 프리 타르트 반죽에 탄력과 응집력을 더해주어 구워진 타르트 셸이 쉽게 무너지지 않도록 도와줍니다.

슈거파우더

타르트 반죽은 다른 제과 반죽에 비해 수분 함량이 적기 때문에 설탕을 사용할 경우 설탕이 반죽 안에서 충분히 녹지 못할 수 있습니다. 설탕 입자가 남아 있을 경우 구워진 타르트 셸 표면에 설탕 결정이 남게 되고, 수분에 녹지 못한 설탕이 캐러멜화되어 보슬보슬한 식감을 내게 됩니다. 따라서 입자가 작은 슈거파우더를 사용하면 반죽 속에서 더 쉽게 분산되며, 표면이 매끄럽고 입안에서 바스러지는 식감을 낼 수 있습니다.

베이킹파우더

글루텐 프리 파이 반죽에 베이킹파우더를 첨가하면 반죽이 잘 부풀어 부서지는 식감을 한층 더 살려줄 수 있습니다. 베이킹파우더는 수분과 열에 의해 반응하기 때문에 밀봉하여 서늘한 곳에 보관하는 것이 좋으며, 개봉 후 너무 오래된 경우 베이킹파우더의 주 성분인 탄산수소나트륨과 산성제 성분이 상온에서 서로 반응하여 팽창 능력이 떨어지므로 너무 오래된 베이킹파우더는 사용하지 않는 것이 좋습니다.

드라이버터

타르트 반죽을 얇은 두께로 성형하거나 작업실 온도가 비교적 높은 경우 작업성을 좋게 하기 위해 일반 버터에 비해 가소성 범위가 넓고 융점이 높은 드라이버터를 사용할 수 있습니다.

세몰리나

세몰리나는 듀럼밀을 부순 가루입니다. 파스타, 시리얼, 푸딩, 쿠스쿠스 등을 만드는 데 사용합니다. 세몰리나는 수분 흡수가 더뎌 사블레의 바삭한 식감을 오랫동안 유지할 수 있게 해주며 거친 입자가 씹히는 식감을 더해줍니다. 입자가 클수록 씹히는 식감이 커지기 때문에 취향에 따라 원하는 입자의 세몰리나를 선택해 사용할 수 있습니다.

Tapioca starch

Tapioca starch adds elasticity and cohesion to the gluten-free tarte dough, helping to prevent the baked tarte shells from collapsing easily.

Powdered sugar

Because tarte dough has less moisture content compared to other confectionery doughs, if sugar is used it may not be able to melt sufficiently in the dough. If sugar particles remain, sugar crystals will remain on the surface of the baked tarte shells, and the sugar that could not melt in water becomes caramelized, which results in brittle texture. Therefore, using the small particles of powdered sugar makes it easier to disperse in the dough, creating a smooth surface and gives crumbly texture in the mouth.

Baking powder

When baking powder is added to gluten-free pie dough, it will help to expand better and give an extra crumbly texture. Baking powder reacts to moisture and heat, so it should be stored airtight in a cool, dry place. If it's been too long after the baking powder has been opened, sodium bicarbonate and acidic agents—the main components—react to each other at room temperature, which reduces the ability to expand. Consequently, it is not recommended to use old baking powder.

Dry butter

In order to improve workability when the dough is rolled thin, or the room temperature of the studio is relatively high, dry butter with a wider range of plasticity and higher melting point can be used other than general butter.

Semolina

Semolina is a powder made by crushing durum wheat. It is used to make pasta, cereals, puddings, couscous, etc. Semolina absorbs moisture slower, which helps to keep the crispy texture of the sablé longer and adding a crunchy bite from the coarse particles. The bigger the size of the particles, the bigger the texture—so one can choose the desired size of semolina particle according to preference.

도구 Tools

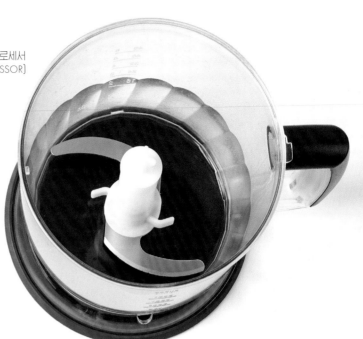

푸드프로세서
[FOOD PROCESSOR]

짤주머니 [PIPING BAG]

저울 [SCALE]

스크래퍼
[SCRAPER]

깍지 [TIPS]

핸드블렌더
[HAND BLENDER]

믹싱볼 [MIXING BOWLS]

스패출러 [SPATULA]

저울

베이킹에 있어 정확한 계량은 매우 중요합니다. 1g 차이로도 맛이 달라지거나 제품이 아예 만들어지지 않을 수도 있기 때문입니다. 따라서 눈금저울보다는 1g 단위로 표시되는 전자저울을 사용하는 것이 좋습니다. 재료를 담는 볼을 올리고 0을 맞춰 0점 조절을 한 다음 계량을 시작합니다. 1g보다 더 적은 양을 계량한다면 1g에서 1/2, 1/4과 같이 줄여나가면 됩니다.

푸드프로세서

재료를 잘게 자르고 다지는 용도로 사용합니다.

짤주머니/깍지

반죽이나 크림을 담아 짜거나 모양을 내기 위해 사용하는 도구입니다. 굽는 반죽의 경우 천 재질의 짤주머니를 사용해도 좋지만, 가열하는 과정 없이 바로 먹는 크림 종류의 경우에는 식중독 예방을 위해 일회용 비닐 짤주머니를 사용하고 재사용하지 않는 것이 위생적입니다. 다양한 종류의 깍지를 짤주머니에 끼워 용도에 맞게 사용합니다.

믹싱볼

반죽을 섞거나 거품을 올리는 작업에 사용되는 필수 도구입니다. 스테인리스, 유리, 폴리카보네이트 등 다양한 재질이 있으며 용도에 맞춰 큰 사이즈(지름 25cm)부터 작은 사이즈(17cm)까지 3~4개 정도 여유롭게 준비하는 것이 좋습니다. 스테인리스 재질은 가볍고 열 전달이 빠른 반면, 유리와 폴리카보네이트 재질은 상대적으로 열 전달이 느려 뜨거운 것이나 차가운 것을 보다 오래 유지시켜줍니다.

핸드블렌더

글레이즈, 크림, 가나슈 등을 균일하게 혼합하는 데 유용한 도구입니다.

스패출러

반죽이나 크림을 펼치거나 고르게 정리할 때 사용합니다. 손잡이와 날이 일직선인 것, L자로 굽은 것 두 가지 종류가 있습니다.

스크래퍼

넓은 면적을 이용해 반죽을 이기거나 혼합하는 데 사용하며, 반죽을 떠 담기에도 유용한 도구입니다.

Scale

Accurate scaling is extremely important in baking. It's because the taste may change, or the product may not be made at all, even with a 1 gram of difference. Therefore, it is better to use an electronic scale that displays in 1-gram increments rather than a needle-type scale. Place a bowl for scaling the ingredient on the scale and set to 0 (tare), and start weighing. If you measure less than 1 gram, reduce from 1 gram to 1/2, 1/4, and so on.

Food processor

This is used to chop and grind ingredients.

Piping bags/tips (nozzles)

These are used for piping or shaping batter or cream. Cloth material piping bags may be used for batters that will be baked. But for creams consumed without the cooking process, it is recommended to use disposable plastic piping bags for hygienic reasons and not to re-use them to prevent food poisoning. Use various types of tips with piping bags accordingly.

Mixing bowls

It is an indispensable tool used for mixing doughs or whipping. There are various materials such as stainless steel, glass, polycarbonate, etc. It is recommended to prepare 3~4 bowls, from large size (25cm diameter) to small size (17cm diameter), to use according to the purpose. Stainless steel material is light and transfers heat faster, while glass and polycarbonate materials have relatively slower heat transfer, which helps keep hots or colds longer.

Hand blender

A hand blender is a useful tool for mixing glazes, creams, and ganache evenly.

Spatula

It is used to spread or evenly organize cream or batter. There are two types: one that's straight from the handle to the blade and one in which the blade is bent to L-shape.

Scraper

A scraper is used to beat or mix batters and is also a useful tool for transferring batter.

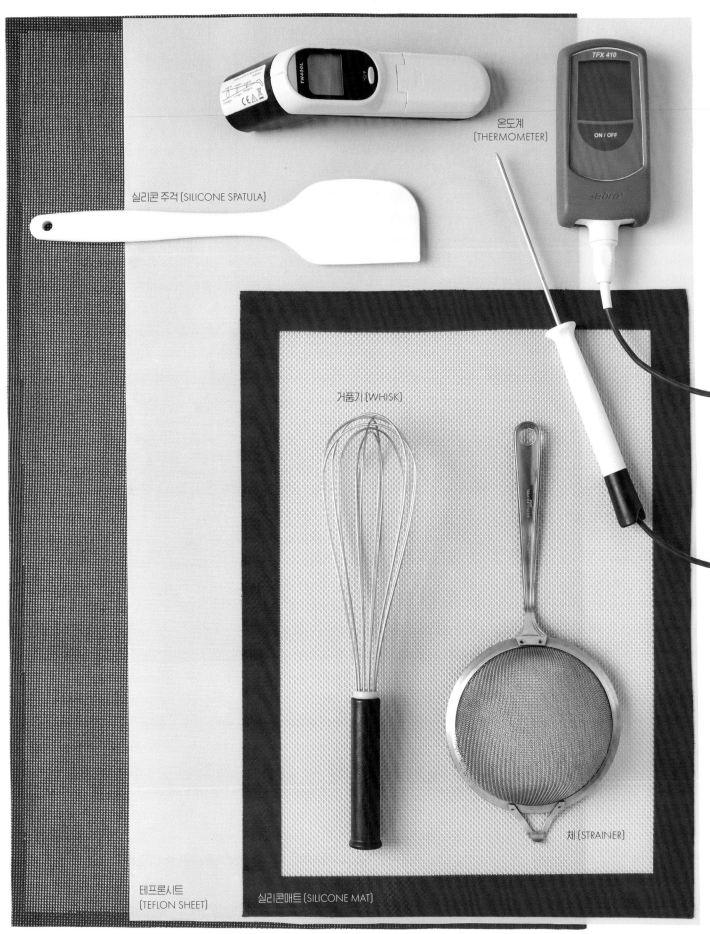

온도계 [THERMOMETER]

실리콘 주걱 [SILICONE SPATULA]

거품기 [WHISK]

체 [STRAINER]

테프론시트 [TEFLON SHEET]

실리콘매트 [SILICONE MAT]

에어매트 [AIR MAT]

오븐용 페이퍼

오븐용 페이퍼는 유산지, 테프론시트, 실리콘매트 등의 제품을 말하며 철판에 깔아 제품과 철판이 달라붙는 것을 방지하고 구웠을 때 제품의 밑면이 깔끔하게 분리되게 하는 역할을 합니다. 실리콘 페이퍼, 내열 페이퍼, 방수 페이퍼 등의 특수 가공 처리를 한 제품들도 있습니다.

에어매트

그물 모양으로 만들어진 오븐용 실리콘매트입니다. 촘촘하게 구멍이 뚫려 있기 때문에 오븐 안에서 공기의 흐름이 좋고 반죽이 미끄러지지 않아 모양이 그대로 유지됩니다.

온도계

정확한 계량 못지않게 베이킹에서 중요한 것이 바로 온도입니다. 온도계는 주로 반죽이나 크림의 온도를 체크하는 데 사용되며 고온의 시럽을 만들 때에도 사용할 수 있도록 250℃까지 측정이 되는 제품을 선택하는 것이 좋습니다. 표면의 온도를 측정하는 비접촉식 적외선 온도계와 내용물 안에 넣어 온도를 측정하는 접촉식 온도계 두 가지 종류가 있습니다.

실리콘 주걱

볼에 담긴 반죽이나 크림을 섞거나 모으는 데 사용합니다. 힘 없이 휘어지는 것보다는 어느 정도 탄력이 있는 제품이 사용하기 더 편리합니다. 주걱 앞부분부터 손잡이 끝부분까지 홈이 없이 매끄럽게 연결되어 만들어진 주걱이 이물질이 끼지 않아 위생적으로 더 좋습니다.

거품기

재료를 섞어 혼합하거나 공기를 포집할 때 사용하는 도구입니다. 와이어 부분이 유연하면서도 튼튼하며 손으로 잡았을 때 편한 것을 고르는 것이 좋습니다.

체

가루 재료를 곱게 내리거나 섞어 놓은 액체 재료를 거르는 데 사용하는 도구로 다양한 사이즈가 있습니다. 제과에서 가루 재료들은 모두 체에 내린 후 사용하는 것이 기본입니다. 가루 재료를 체에 내리면 뭉쳐 있던 멍울은 풀어지고, 불순물이 제거되며, 체에 내리는 과정에서 가루와 가루 사이에 공기가 혼입되어 다른 재료와 더 잘 섞이게 됩니다.

Papers for oven

Papers for the oven refers to parchment paper, Teflon sheets, and silicone mats. These are to be lined on a baking tray to prevent products from sticking and to help the bottom of the baked goods separate neatly. There are also specially processed products such as silicone paper, heat-resistant paper, and waterproof paper.

Air mat

This is a net-shaped silicone mat for ovens. Because of its fine holes, the air in the oven flows through and around, and the dough does not slip on it, which helps to maintain its shape.

Thermometer

Temperature is every bit as important as accurate scaling when it comes to baking. Thermometers are mainly used to check the temperature of dough/batter or cream and check the temperature of high-temperature syrups. Therefore, it is recommended to select a product that can measure up to 250℃. There are two types: a non-contact type infrared thermometer that measures surface temperature and a contact type thermometer that measures the temperature by placing it inside the content.

Silicone Spatulas

These are used to mix or collect batter or cream in the bowl. Spatula with a resilient blade is more convenient to use than the blades that bend easily. The spatula that is smooth all the way, and has no groove from the blade to the tip of the handle, is better in the means of hygiene because no foreign substance will remain in between.

Whisk

This is a tool used to mix ingredients or to collect air. Choosing a whisk with wires that are flexible yet sturdy and comfortable when held by hand is good to use.

Strainer

A strainer is used to sift powder ingredients a fine form or filter mixture of liquid ingredients, and there are various sizes available. In confectionery, it is a ground rule to sieve all powdered ingredients. When the powdered ingredients are sifted, the lumps break up, impurities are removed, and the air gets to mix in between the particles, which helps to better mix with other ingredients.

타르트 틀

Tarte Mold

타르트 틀은 바닥과 테두리가 분리되는 '분리형 틀', 바닥과 테두리가 일체된 '일체형 틀', 바닥 없이 테두리만 있는 '링 모양 틀' 등 다양한 타입이 있으며, 원형, 삼각형, 사각형, 국화 형태 등 모양 또한 다양해 취향이나 사용 목적에 따라 선택해 사용할 수 있습니다.

타공된 타르트 틀의 경우 오븐 안에서 공기의 흐름이 좋아 반죽이 고르게 구워지며, 반죽의 수증기 배출도 원활해 누름돌을 얹지 않아도 굽는 도중 반죽이 부분적으로 부풀어버리거나 들뜨는 현상이 없습니다. 또한 타공된 부분에 의해 반죽이 미끄러지지 않아 흘러내림 현상이 적습니다.

타공이 되지 않은 일반적인 형태의 타르트 틀의 경우 반죽에 피케(반죽 표면에 포크와 같이 뾰족한 도구로 구멍을 내는 것) 작업을 하여 반죽의 수증기가 원활하게 배출되도록 하는 것이 좋습니다. 또한 반죽이 부풀어버리거나 들뜨는 것을 방지하기 위해 누름돌을 얹고 구워내는 것이 좋습니다.

There are various types of tarte molds; 'removable mold', which the bottom and the frame separates, 'integral mold', which the bottom and the frame is one single unit, 'ring mold', which is a ring frame without a bottom. Also, various shapes are available that can be chosen according to preference or purpose, such as round, triangle, square, chrysanthemum shape, etc.

In the case of a perforated mold, the dough is baked evenly due to the good airflow in the oven, and helps to release moisture from the dough smoothly, so there is no partial bubbling or lifting of the dough during baking even without the use of pie weights. Also, because of the perforations, the dough does not slide, so it does not run down easily.

When using regular tarte molds that are not perforated, the dough should be docked (a process of poking holes on the dough with pointy tools such as forks), so the steam discharges smoothly from the dough. Also, it helps to bake with pie weights to prevent the dough from inflating or lifting away from the mold.

글루텐 프리 레시피를 위한 재료들

Ingredients for Gluten-free Recipes

디저트 샵을 운영하다보면 알레르기로 인해 글루텐 섭취가 어려운 분들을 종종 만나게 됩니다. 이런 분들을 보며 팀 가루하루의 호기심과 고민
이 시작되었고, 수많은 테스트를 거쳐 지금의 글루텐 프리 레시피를 완성할 수 있었습니다.
이 책에서 소개하는 글루텐 프리 레시피를 활용하면 글루텐 알레르기가 있으신 분들은 물론 건강상의 이유로 글루텐 섭취를 제한하고 싶은 분
들까지도 맛있고 건강한 가루하루의 타르트를 즐기실 수 있습니다.

○ 쌀가루

**시판되는
쌀가루의 종류**

쌀가루는 침지(불림) 과정의 유무에 따라 '건식쌀가루'와 '습식쌀가루'로 구분되며 각각의 제조 과정은 아래와 같습
니다.

• 건식쌀가루의 제조 과정

세척	탈수	제분	건조

• 습식쌀가루의 제조 과정

세척	침지(불림)	탈수	제분	건조

침지 과정은 쌀 원물을 물에 불려 전분 입자 사이의 간격을 벌어지게 해 스펀지와 같은 조직을 만드는 과정입니다.
상온에서 유통되는 습식 쌀가루의 경우 이렇게 쌀 전분 입자를 벌려 놓은 상태에서 건조시키므로(상온 유통 과정 중
미생물이 번식하는 것을 방지하기 위해) 순간적으로 많은 양의 수분을 흡수할 수 있게 됩니다.
제과에서 밀가루 대체 재료로 쌀가루를 사용할 경우에는 수분 흡수와 보유력이 좋은 습식쌀가루를 사용하는 것이 적
합합니다. 건식쌀가루의 경우 반죽 과정에서 쌀 전분이 충분한 수분을 흡수할 수 없어 반죽이 질어지고 퍼지는 요인
이 되기 때문입니다. 또한 입자가 고울수록 부드러운 식감을 주기 때문에 기호나 용도에 따라 쌀가루의 입자 크기를
선택해 사용할 수 있습니다. 시판되고 있는 박력쌀가루의 경우 부드러운 식감을 위해 밀가루와 비슷한 입자로 제분
한 것입니다.

When you run a dessert shop, you will often meet people who have difficulty ingesting gluten due to allergies. As I watched these people, Team GARUHARU's curiosity and concerns began, and after numerous tests, we were able to complete the current gluten-free recipe.

By using the gluten-free recipe introduced in this book, you can enjoy GARUHARU's delicious and healthy tarte, not only for those with gluten allergies but also for those who want to limit gluten intake for health reasons. We hope you will enjoy the various styles of tartes by choosing recipes presented in this book.

○ Rice flour

Types of rice flour available in the market

Rice flour is divided into 'dry(-milled) rice flour' and 'wet(-milled) rice flour' depending on whether or not it went through the process of immersion (soaking), and each manufacturing process is as follows.

• Manufacturing process of dry rice flour

| Wash | Dehydration | Milling | Dry |

• Manufacturing process of wet rice flour

| Wash | Immersion (soaking) | Dehydration | Milling | Dry |

The immersion is a process in which raw rice grains are soaked in water to open the gaps between starch particles to create a sponge-like tissue. In the case of the wet rice flours distributed at ambient temperature, rice starch particles are dried in an open state (to prevent microorganisms from multiplying during the distribution process in ambient temperature) so that a large amount of moisture can be absorbed instantaneously.

When using rice flour to substitute wheat flour in confectionery, it is suitable to use wet rice flour, which has good moisture absorption and retention. This is because, when dry rice flour is used, the rice starch cannot absorb enough moisture during the kneading process, which can cause the dough to become thin and run easily. Additionally, the finer the particles, the softer the texture, so you can select the particle size of rice flour and use it according to your taste or purpose. The commercially available soft rice flour is ground into particles similar to wheat flour for a smooth texture.

쌀가루 만들기

1. 쌀을 흐르는 물에 3~4회 깨끗이 세척한다.

2. 냉장고에서 12시간 정도 불린다.

3. 체에 받쳐 30분간 탈수한다.

4. 45℃에서 10시간 정도 건조시킨다.

5. 푸드프로세서에 넣고 고운 가루 상태가 될 때까지 갈아준다.

How to make rice flour

1. Wash rice 3~4 times under running water.

2. Let it soak in refrigerator for about 12 hours.

3. Strain and dehydrate for 30 minutes.

4. Dry at 45℃ for 10 hours.

5. Grind in food processor until it turns into fine powder.

파트 아 사블레의 이해

Understanding PATE A SABLÉ

사블레 반죽 버터와 달걀 풍미가 풍부한 사블레 반죽은 쉽게 부서지는 특징을 가지고 있습니다. '사블레'는 '모래(sable)'라는 뜻을 가지고 있는데 구워낸 반죽이 입안에서 모래처럼 부서진다고 해서 붙여진 이름입니다.

밀가루와 버터를 사블라주*하는 방식으로 만드는 사블레 반죽은 밀가루와 수분의 연결이 좋지 않아 글루텐 형성을 막게 되며 이 때문에 잘 부서지고 바삭한 식감으로 완성됩니다.

사블라주* 사블라주는 유지와 밀가루를 반죽하지 않고 가볍게 섞어 모래알 상태로 만드는 것을 말합니다. 이렇게 작업하면 밀가루 입자에 버터 유지막이 형성되고, 이로 인해 다음 과정에서 넣는 수분과 밀가루의 연결이 어려워져 글루텐 생성이 최소화됩니다.

버터가 무르면 사블라주 작업 중 밀가루가 버터의 수분을 흡수해 반죽이 경단 상태로 뭉쳐버리므로 가루에 붙은 유지의 막이 고르지 않게 됩니다. 따라서 재료는 반드시 차가운 상태로 준비해야 합니다.

PATE A SABLÉ Rich in butter and egg flavor, the sablé dough has the characteristic of breaking easily. 'Sablé' means 'sand (sable)'; it was named so because the baked dough crumbles like sand when consumed. Sablé dough, which requires sablage* of flour and butter, lacks a connection between the flour and the water, which prevents the formation of gluten, and because of this reason, it makes a crumbly and crispy texture.

SABLAGE* Sablage refers to lightly mixed wheat flour and milk fat without kneading to make it into a sand-like structure. This method creates butter fat membranes on the flour particles, and as a result, the water added in the following step cannot connect with the flour, minimizing the formation of gluten.

If the butter is soft, the flour absorbs the moisture from the butter, which will cause the dough to turn into smooth lumpy balls, creating uneven layers of butter adhered to the flour. Therefore, the ingredients must be prepared cold.

파트 아 브리제의 이해
Understanding PATE A BRISÉE

브리제 반죽

브리제 반죽은 밀가루와 물을 기본으로 한 반죽에 버터가 분산되어 있는 구조입니다. 설탕을 넣지 않아 단맛이 없고 담백한 것이 특징입니다.

밀가루와 버터를 사블라주(33p)*한 후 수분을 넣어 글루텐 형성을 최소화한 브리제 반죽은 사각사각 부드럽게 부서지는 식감으로 완성됩니다.

PATE A BRISÉE

Brisée dough is a structure of dispersed butter in a dough based on flour and water. It is light and has no sweetness because it does not contain sugar. Brisée dough, in which gluten formation is limited by adding the water after the sablage of flour and butter (33p)*, gives a crispy texture that crumbles softly.

컨벅션 오븐과 데크 오븐
Convection Oven and Deck Oven

오븐의 종류는 '데크 오븐'과 '컨벡션 오븐'으로 나눌 수 있습니다.

데크 오븐은 오븐의 상부와 하부에 위치한 열선을 통해 열을 전달하며, 컨벡션 오븐은 오븐 내부의 팬이 오븐 안의 열을 순환시키는 방식으로 열을 전달합니다.

컨벡션 오븐의 경우 뜨거운 바람으로 열을 전달하기 때문에 다단 조리가 가능한 장점이 있지만 반죽이 건조해지기 쉽습니다. 따라서 컨벡션 오븐은 바삭한 식감의 쿠키나 파이 등을 굽기에 적합하며, 촉촉한 식감의 제품을 구울 때는 적합하지 않습니다.

데크 오븐 DECK OVEN

상부와 하부의 열선을 통해 열을 전달

Heat distributed through
the upper and lower heating wires

컨벡션 오븐 CONVECTION OVEN

팬의 바람을 통해 열을 순환

Distributes heat by circulating heated air
with a fan

The types of ovens can be divided into 'deck oven' and 'convection oven.'

The deck oven distributes heat through the heating wires located at the top and bottom of the oven. As for the convection oven, heated air is distributed by the fan inside the oven circulating the heat.

Convection ovens have the advantage of being able to cook multiple trays because it transfers heat by hot air, but the dough may easily dry out. Therefore, a convection oven is ideal for baking crispy textured cookies or pies and not suitable for baking products that need to be moist.

Basic

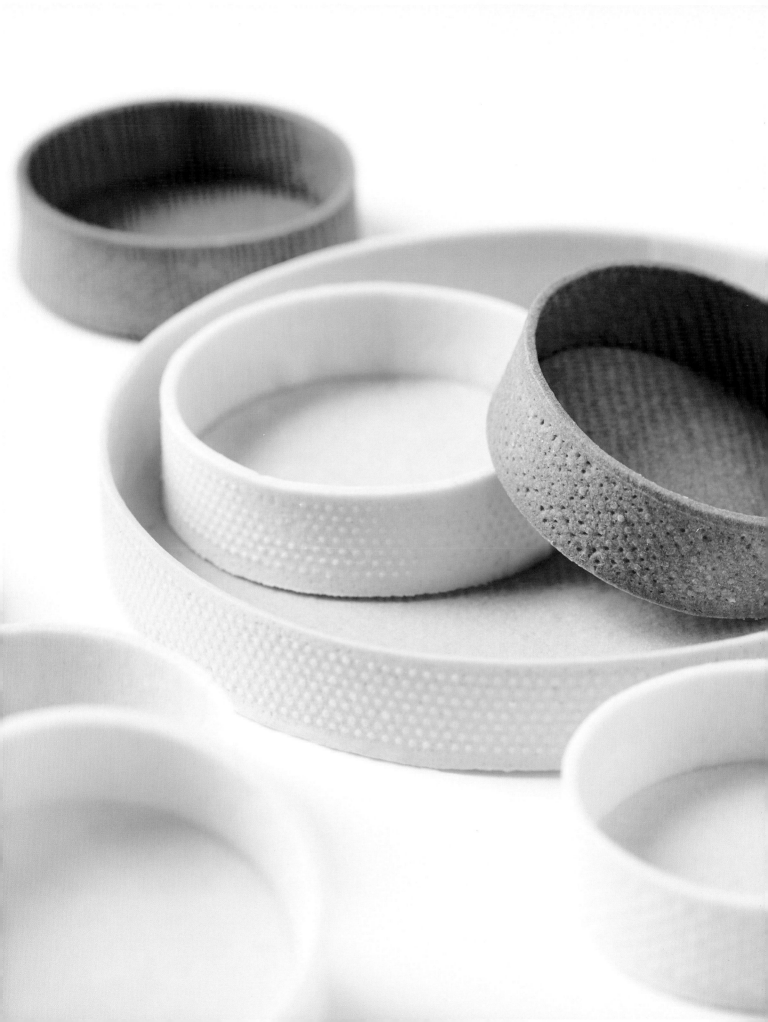

¹ PATE A SABLÉ

파트 아 사블레

Ø 7cm
15ea

CONVECTION OVEN
160°C

ingredients

드라이버터 86g

중력분 163g

슈거파우더 82g

소금 0.9g

달걀전란 48g

아몬드파우더 12g

세몰리나(중간 굵기) 12g

전분 47g

86g Dry butter

163g All-purpose flour

82g Powdered sugar

0.9g Salt

48g Whole eggs

12g Almond powder

12g Semolina (medium grind)

47g Starch

Process

1. 모든 재료는 차가운 상태로 준비한다.

2. 푸드프로세서에 중력분, 사방 2cm 크기로 자른 드라이버터를 넣는다.

3. 모래알 상태가 될 때까지 믹싱한다.

4. 나머지 재료를 모두 넣고 반죽이 어느 정도 뭉쳐질 때까지 계속해서 믹싱한다.

5. 반죽을 작업대 위에 올린 다음 한 덩어리가 되도록 가볍게 뭉쳐준다.

6. 스크래퍼를 이용해 반죽을 밀어내듯 펴주면서 재료를 고르게 혼합한다.

7. 완성된 반죽은 마르지 않도록 랩으로 감싼 후 냉장고에서 최소 3시간 휴지시킨다.

8. 1.5mm 두께로 밀어 편 후 원하는 사이즈로 성형해 오븐에서 굽는다. (69p)

1. All ingredients should be prepared cold.

2. In a food processor, add all-purpose flour and dry butter cut to 2cm cubes.

3. Grind until the mixture resembles coarse sand.

4. Add the rest of the ingredients and continue mixing until the dough becomes somewhat homogeneous.

5. Place the dough on the worktable and lightly knead to form a ball.

6. Using a scraper, spread the dough as if to push away, to mix the ingredients evenly.

7. Wrap the finished dough with plastic wrap to prevent it from drying, and store in the fridge to rest at least 3 hours.

8. Roll out to 1.5mm, and shape to the desired size to bake. (69p)

² PATE A SABLÉ
- CHOCOLATE

파트 아 사블레 - 초콜릿

Ø 7cm
15ea

CONVECTION OVEN
❀ 160°C

ingredients

드라이버터 86g

중력분 154g

슈거파우더 82g

소금 0.9g

달걀전란 48g

아몬드파우더 12g

세몰리나(중간 굵기) 12g

카카오파우더 17g

전분 39g

86g Dry butter

154g All-purpose flour

82g Powdered sugar

0.9g Salt

48g Whole eggs

12g Almond powder

12g Semolina (medium grind)

17g Cacao powder

39g Starch

Process

1. 모든 재료는 차가운 상태로 준비한다.

2. 푸드프로세서에 중력분, 사방 2cm 크기로 자른 드라이버터를 넣는다.

3. 모래알 상태가 될 때까지 믹싱한다.

4. 나머지 재료를 모두 넣고 반죽이 어느 정도 뭉쳐질 때까지 계속해서 믹싱한다.

5. 반죽을 작업대 위에 올린 다음 한 덩어리가 되도록 뭉쳐준다.

6. 스크래퍼를 이용해 반죽을 밀어내듯 펴주면서 재료를 고르게 혼합한다.

7. 완성된 반죽은 마르지 않도록 랩으로 감싼 후 냉장고에서 최소 3시간 휴지시킨다.

8. 1.5mm 두께로 밀어 편 후 원하는 사이즈로 성형해 오븐에서 굽는다. (69p)

1. All ingredients should be prepared cold.

2. In a food processor, add all-purpose flour and dry butter cut to 2cm cubes.

3. Grind until the mixture resembles coarse sand.

4. Add the rest of the ingredients and continue mixing until the dough becomes somewhat homogeneous.

5. Place the dough on the worktable and knead to form a ball.

6. Using a scraper, spread the dough as if to push away, to mix the ingredients evenly.

7. Wrap the finished dough with plastic wrap to prevent it from drying, and store in the fridge to rest at least 3 hours.

8. Roll out to 1.5mm, and shape to the desired size to bake. (69p)

³ PATE A SABLÉ
- MATCHA

파트 아 사블레 – 말차

Ø 7cm
18ea

CONVECTION OVEN
⊛ 160℃

ingredients

드라이버터 87g

중력분 150g

슈거파우더 82g

소금 1.4g

달걀전란 49g

아몬드파우더 12g

세몰리나 12g

말차파우더 17g

전분 40g

87g Dry butter

150g All-purpose flour

82g Powdered sugar

1.4g Salt

49g Whole eggs

12g Almond powder

12g Semolina

17g Matcha powder

40g Starch

Process

1.   모든 재료는 차가운 상태로 준비한다.

2.   푸드프로세서에 중력분, 사방 2cm 크기로 자른 드라이버터를 넣는다.

3.   모래알 상태가 될 때까지 믹싱한다.

4.   나머지 재료를 모두 넣고 반죽이 어느 정도 뭉쳐질 때까지 계속해서 믹싱한다.

5.   반죽을 작업대 위에 올린 다음 한 덩어리가 되도록 뭉쳐준다.

6.   스크래퍼를 이용해 반죽을 밀어내듯 펴주면서 재료를 고르게 혼합한다.

7.   완성된 반죽은 마르지 않도록 랩으로 감싼 후 냉장고에서 최소 3시간 휴지시킨다.

8.   1.5mm 두께로 밀어 편 후 원하는 사이즈로 성형해 오븐에서 굽는다. (69p)

1.   All ingredients should be prepared cold.

2.   In a food processor, add all-purpose flour and dry butter cut to 2cm cubes.

3.   Grind until the mixture resembles coarse sand.

4.   Add the rest of the ingredients and continue mixing until the dough becomes somewhat homogeneous.

5.   Place the dough on the worktable and knead to form a ball.

6.   Using a scraper, spread the dough as if to push away, to mix the ingredients evenly.

7.   Wrap the finished dough with plastic wrap to prevent it from drying, and store in the fridge to rest at least 3 hours.

8.   Roll out to 1.5mm, and shape to the desired size to bake. (69p)

4 SWEET DOUGH
- GLUTEN FREE

스위트 반죽 - 글루텐 프리

7cm
12ea

CONVECTION OVEN
❄160℃

ingredients

드라이버터 109g	109g Dry butter
쌀가루 177g	177g Rice flour
아몬드파우더 33g	33g Almond powder
타피오카 전분 15g	15g Tapioca starch
슈거파우더 70g	70g Powdered sugar
소금 0.9g	0.9g Salt
달걀전란 61g	61g Whole eggs

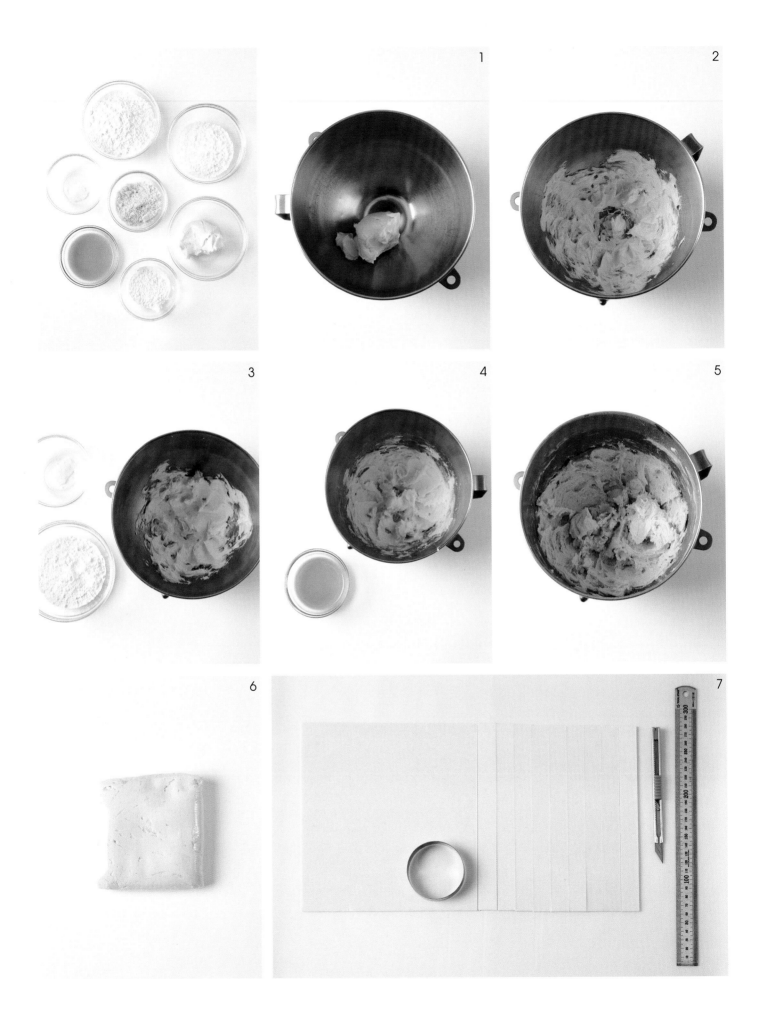

Process

1. 드라이버터는 상온 상태로 준비한다.

2. 드라이버터를 부드럽게 풀어준다.

3. 슈거파우더, 소금을 넣고 혼합한다.

4. 달걀전란을 조금씩 나눠 넣어가며 섞는다.

5. 쌀가루, 아몬드파우더, 타피오카 전분을 넣고 섞는다.

6. 완성된 반죽은 마르지 않도록 랩으로 감싼 후 냉장고에서 최소 3시간 휴지시킨다.

7. 2mm 두께로 밀어 편 후 원하는 사이즈로 성형해 오븐에서 굽는다. (69p)
    * 글루텐 프리 반죽은 응집력이 떨어져 굽고난 후 쉽게 무너지기 때문에 일반적인 타르트 반죽보다 조금 더 두껍게
      성형하는 것이 좋으며, 완전히 식힌 후에 틀에서 빼내는 것이 좋다.

1. Prepare dry butter at room temperature.

2. Beat to soften the dry butter.

3. Mix with powdered sugar and salt.

4. Gradually add whole eggs while mixing.

5. Mix with rice flour, almond powder, and tapioca starch.

6. Wrap the finished dough with plastic wrap to prevent it from drying, and store
    in the fridge to rest at least 3 hours.

7. Roll out to 2mm, and shape to the desired size to bake. (69p)
    * Gluten free dough collapses easily after baking due to lack of cohesiveness, so it is better to roll
      out a little bit thicker than regular tarte dough. Also, it is recommended to cool completely before
      removing from the molds.

^{5} PATE A BRISÉE

파트 아 브리제

Ø 7cm
12ea

CONVECTION OVEN
175°C

ingredients

박력분 244g

드라이버터 122g

달걀노른자 32g

물 49g

소금 1.8g

244g Cake flour

122g Dry butter

32g Egg yolks

49g Water

1.8g Salt

Process

1. 모든 재료는 차가운 상태로 준비한다.

2. 물, 달걀노른자, 소금은 미리 혼합한다.

3. 푸드프로세서에 박력분, 사방 2cm 크기로 자른 드라이버터를 넣고 보슬보슬한 상태가 될 때까지 믹싱한다.

4. **2**를 넣고 날가루가 보이지 않을 때까지 계속해서 믹싱한다.

5. 반죽을 작업대 위에 올린 다음 한 덩어리가 되도록 뭉쳐준다.

6. 반죽을 반으로 자른 후 포개어준다.

7. **5-6** 과정을 3~4회 반복한다.

8. 완성된 반죽은 마르지 않도록 랩으로 감싼 후 냉장고에서 최소 3시간 휴지시킨다.

9. 2mm 두께로 밀어 편 후 원하는 사이즈로 성형해 오븐에서 굽는다. (69p)

1. All ingredients should be prepared cold.

2. Mix ahead water, egg yolks, and salt.

3. In a food processor, add cake flour and dry butter cut to 2cm cubes and mix until the mixture resembles coarse sand.

4. Continue to mix with **2** until no dry ingredients are visible.

5. Place the dough on the worktable and knead to form a ball.

6. Cut the dough in half and stack one piece on top of the other.

7. Repeat steps **5-6**, 3~4 times.

8. Wrap the finished dough with plastic wrap to prevent it from drying, and store in the fridge to rest at least 3 hours.

9. Roll out to 2mm, and shape to the desired size to bake. (69p)

GLUTEN
FREE

⁶ PIE DOUGH
- GLUTEN FREE

파이 반죽 – 글루텐 프리

Ø 7cm
12ea

CONVECTION OVEN
❀ 175°C

ingredients

쌀가루 199g

타피오카 전분 18g

베이킹파우더(알루미늄 프리 & 글루텐 프리) 0.5g

드라이버터 122g

달걀전란 109g

소금 2g

199g Rice flour

18g Tapioca starch

0.5g Baking powder (aluminum free & gluten free)

122g Dry butter

109g Whole eggs

2g Salt

Process

1. 모든 재료는 차가운 상태로 준비한다.

2. 달걀전란, 소금은 미리 혼합한다.

3. 푸드프로세서에 쌀가루, 타피오카 전분, 베이킹파우더, 사방 2cm 크기로 자른 차가운 상태의 드라이버터를 넣는다.

4. 보슬보슬한 상태가 될 때까지 믹싱한다.

5. **2**를 넣고 날가루가 보이지 않을 때까지 계속해서 믹싱한다.

6. 반죽을 작업대 위에 올린 다음 한 덩어리가 되도록 뭉쳐준다.

7. 반죽을 반으로 자른 후 포개어준다.

8. **6-7** 과정을 3~4회 반복한다.

9. 완성된 반죽은 마르지 않도록 랩으로 감싼 후 냉장고에서 최소 3시간 휴지시킨다.

10. 2.5mm 두께로 밀어 편 후 원하는 사이즈로 성형해 오븐에서 굽는다. (69p)

1. All ingredients should be prepared cold.

2. Mix ahead whole eggs and salt.

3. In a food processor, add rice flour, tapioca starch, baking powder, and cold, dry butter cut to 2cm cubes.

4. Mix until the mixture resembles coarse sand.

5. Mix with **2** until no dry ingredients are visible.

6. Place the dough on the worktable and knead to form a ball.

7. Cut the dough in half and stack one piece on top of the other.

8. Repeat steps **6-7**, 3~4 times.

9. Wrap the finished dough with plastic wrap to prevent it from drying, and store in the fridge to rest at least 3 hours.

10. Roll out to 2.5mm, and shape to the desired size to bake. (69p)

라이트 아몬드 크림
(LIGHT ALMOND CREAM)

클래식 아몬드 크림
(CLASSIC ALMOND CREAM)

⁷ ALMOND CREAM

아몬드 크림

아몬드 크림은 포마드 상태의 버터에 슈거파우더, 아몬드파우더, 달걀을 혼합한 크림으로 주로 타르트 또는 파이 반죽에 채워 굽는 형태로 사용합니다.

아몬드 크림의 기본 배합은 버터, 슈거파우더, 아몬드파우더, 달걀이 동량입니다. 본 책에서는 메뉴에 따라 이 기본 레시피를 약간 변형한 '클래식 아몬드 크림', 버터 양을 줄이고 머랭을 더해 볼륨감을 준 가루하루 스타일의 '라이트 아몬드 크림'을 사용했습니다.

클래식 아몬드 크림은 부드럽고 묵직한 아몬드 크림 특유의 질감을, 라이트 아몬드 크림은 가볍고 폭신폭신한 질감을 느낄 수 있습니다.

Almond cream is a mixture of beurre pommade (softened butter) mixed with powdered sugar, almond powder and eggs, and is usually filled in tartes or pies and baked.

The basic recipe for almond cream is equal amounts of butter, powdered sugar, almond powder and eggs. In this book, 'classic almond cream', slightly modified from this basic recipe; GARUHARU style of 'light almond cream', where the amount of butter is reduced, and meringue is used to give volume; are used depending on the menu.

Classic almond cream has a soft and rich texture that is characteristic of almond cream; light almond cream has a light and fluffy texture.

⁸ DOUGH FORMATION

반죽 성형

ingredients

상온 상태의 버터
타르트 반죽
물

tools

타르트 링
붓

Room temperature butter
Tarte dough
Water

Tarte rings
Brush

Process

1. 타르트 링 안쪽에 상온 상태의 버터를 얇게 바른다.

2. 타르트 링 둘레와 높이에 맞춰 재단한 띠 반죽, 타르트 링 지름에 맞춰 재단한 바닥 반죽을 준비한다.
    * 바닥 반죽은 타르트 링 지름보다 5mm 작게 준비한다.

3. 타르트 링 안쪽에 띠 반죽을 고정시킨다.

4. 링 안쪽 반죽(바닥 반죽과의 접착 면)에 물을 발라준다.

5. 바닥 반죽을 넣어 고정시킨다.

1. Brush a thin layer of room temperature butter on the inside of the tarte ring.

2. Prepare a strip of dough cut according to the circumference and the height of the ring, and bottom dough cut to the diameter of the ring.
    * The bottom dough should be 5mm smaller than the diameter of the tarte ring.

3. Fix the strip on the inside of the tarte ring.

4. Brush water on the inside of the strip (where the bottom dough will adhere).

5. Insert and fasten the bottom dough.

타르트 셸의 굽는 정도는 셸 안에 채워지는 필링의 종류에 따라 결정됩니다.

80% 소성

굽는 타입의 필링이 채워지는 경우 에그 워시 분사 후 옅은 갈색이 날 때까지 80% 정도만 굽습니다.

: 160℃ 오븐에서 15분(50% 소성) – 에그 워시 분사 – 160℃ 오븐에서 5분(80% 소성)

100% 소성

프레시한 타입의 필링이 채워지는 경우 타르트 셸이 황금 갈색이 날 때까지 충분히 굽습니다.

: 160℃ 오븐에서 15분(50% 소성) – 에그 워시 분사 – 160℃ 오븐에서 15분(100% 소성)

50% 소성 후 에그 워시 분사
[50% baked and sprayed with egg wash]

80% 소성
[80% baked]

The degree of baking tarte shells depends on the types of fillings that are filled inside.

80% baked
If baking-type fillings are filled, bake about 80% after spraying egg wash, just until the shells are lightly browned.
: 15 minutes at 160°C (50% baked) - spray egg wash - 5 minutes at 160°C
  (80% baked)

100% baked
If fresh-type fillings are filled, bake sufficiently until the shells are golden brown.
: 15 minutes at 160°C (50% baked) - spray egg wash - 15 minutes at 160°C
  (100% baked)

100% 소성
(100% baked)

EGG WASH

1

2-1

2-2

¹⁰ EGG WASH

에그 워시

에그 워시는 타르트 반죽에 수분이 흡수되는 것을 방지해 바삭한 식감을 오랫동안 유지할 수 있게 해줍니다. 스프레이 건과 붓 모두 사용할 수 있지만 스프레이 건을 사용할 경우 균일한 두께로 도포할 수 있으며 작업성 또한 더 좋습니다.

Egg wash prevents moisture from being absorbed into the tarte dough, allowing it to maintain its crispy texture longer. Both spray gun and brush can be used. However, if a spray gun is used the egg wash can be applied in uniform thickness and has better workability.

ingredients

생크림 75g
달걀노른자 60g

tools

스프레이 건
또는 붓

75g Fresh cream	Spray gun
60g Egg yolks	or Brush

Process

1. 생크림과 달걀노른자가 잘 섞이도록 혼합한다.

2. 완성된 에그 워시는 스프레이 건 또는 붓을 이용해 타르트 셸에 얇게 도포한다.

1. Thoroughly mix fresh cream and egg yolks.

2. Use a spray gun or brush to apply thinly on the tarte shells.

CRYSTAL GLAZE

1

2

3-1

3-2

¹¹ CRYSTAL SPRAY GUN MIXTURE

크리스탈 스프레이 건 믹스처

크리스탈 스프레이 건 믹스처는 크림이 마르지 않도록 보호해주는 역할을 합니다. 모양을 내어 파이핑한 크림은 굴곡이 있기 때문에 글레이즈를 흘려 붓는 방식의 글레이징이 불가능합니다. 이 경우 스프레이 건을 이용해 분사하면 굴곡이 있는 크림 표면에 균일한 두께로 글레이징을 할 수 있습니다.

Crystal spray gun mixture protects the cream from drying out. Piped creams have curvatures, which makes it impossible to apply the method of glazing. In this case, the glaze can be applied in even thickness on curved surfaces when spray guns are used.

ingredients

압솔뤼 크리스탈 글레이즈 150g
물 50g

tools

스프레이 건

150g Absolu Cristal glaze
50g Water

Spray gun

Process

1. 압솔뤼 크리스탈 글레이즈와 물을 준비한다.

2. 냄비에 넣고 가열한다.

3. 압솔뤼 크리스탈 글레이즈가 완전히 녹으면 체에 걸러준 후 80℃에 맞춰 분사한다.

1. Prepare Absolu Cristal glaze and water.

2. Heat in a saucepan.

3. When Absolu Cristal glaze is completely dissolved, sieve and use at 80℃.

GELATIN MASS

1

2-1

2-2

3

¹² GELATIN MASS

젤라틴매스

응고제의 일종인 젤라틴은 무스나 젤리 등을 굳히는 데 사용되는 재료입니다. 이 책에서는 가루 형태의 젤라틴(200 bloom)을 5배의 물과 혼합해 굳힌 후 사용하였습니다.

Gelatine is a kind of coagulant that is used to harden mousse or jelly. For this book, the powder type of gelatin (200 bloom) is mixed with five times of water and hardened to use (gelatin mass).

ingredients

200 Bloom 가루젤라틴 10g
물 50g

10g Powdered gelatin, 200 bloom
50g Water

Process

1. 가루젤라틴과 물을 1:5 비율로 계량한다.

2. 가루젤라틴과 물을 혼합한다.

3. 젤라틴이 굳으면 필요한 만큼 적당한 크기로 잘라 사용한다. 냉장 상태에서 2주, 냉동 상태에서 3개월 동안 보관하며 사용할 수 있다.

1. Scale powdered gelatin and water in the ratio of 1:5.

2. Mix powdered gelatin and water.

3. When the gelatin is set, cut into moderate size to use. It can be stored for about two weeks in a refrigerator and three months in a freezer.

13 CHOCOLATE TEMPERING

초콜릿 템퍼링(접종법)

접종법은 녹인 초콜릿에 안정된 결정 상태의 초콜릿을 넣어 템퍼링하는 방법입니다. 여러 가지 템퍼링 방법 중 작업 방식이 간편해 대량 작업 시에도 비교적 손쉽게 템퍼링할 수 있습니다.

The seeding method is a technique of tempering, where stable crystallized chocolate is added to melted chocolate. This method is simple among various methods of tempering, which also makes it relatively easy to work with a large amount of chocolate.

ingredients

커버추어 초콜릿

couverture chocolate

Process

1. 커버추어 초콜릿을 폴리카보네이트 볼에 담고 녹여준다.
   (다크초콜릿 55~58℃, 밀크초콜릿 45~48℃, 화이트초콜릿 45~48℃)

2. 녹인 초콜릿 양의 30% 정도의 초콜릿을 1에 넣고 골고루 저어준 후 1분간 그대로 둔다.

3. 핸드블렌더를 이용해 초콜릿 덩어리가 남지 않도록 균일하게 믹싱하며 온도를 낮춰준다.
   (다크초콜릿 31~32℃, 밀크초콜릿 29~30℃, 화이트초콜릿 28~29℃)
   * 이때 빠른 속도로 장시간 믹싱하면 마찰열에 의해 온도가 과도하게 올라갈 수 있으므로 주의한다.

4. 스패출러 또는 나이프 끝부분에 템퍼링 테스트를 한 후 사용한다.
   * 23~24℃를 유지한 작업실에서 5분 이내에 얼룩 없이 매끄럽게 초콜릿이 굳으면 템퍼링이 잘된 상태이다.

1. Melt couverture chocolate in a polycarbonate bowl. (Dark chocolate 55~58℃, milk chocolate 45~48℃, white chocolate 45~48℃)

2. Stir in evenly about 30% of the melted chocolate to 1, and let stand for 1 minute.

3. Reduce the temperature of chocolate using a hand blender so that it's mixed evenly and no chocolate chunks are left.
   (Dark chocolate 31~32℃, milk chocolate 29~30℃, white chocolate 28~29℃)
   * Keep in mind when mixing at high speed for a long time, the temperature can excessively increase due to heat caused by friction.

4. Test on the tip of a spatula or knife before use.
   * If the chocolate hardens smoothly within 5 minutes in a studio that maintained 23~24℃, the tempering is done properly.

14 CORNET

코르네

삼각형 모양으로 준비한 베이킹 페이퍼를 고깔 모양으로 말아준 후 끝부분을 안쪽으로 접어 풀리지 않도록 고정시켜 사용합니다.

Prepare baking paper cut into triangle, and roll into cone shape. Fold the end inward to fasten, so it won't undo itself.

Fruits

YUJA BLOSSOM TARTE

RASPBERRY BLOSSOM TARTE

APRICOT TARTE

RASPBERRY & PASSION FRUIT TARTE

TROPICAL TARTE

PEAR & VANILLA CREAM TARTE

GRAPEFRUIT TARTE

¹ YUJA BLOSSOM TARTE

유자 블로섬 타르트

ingredients - ⌀ 8cm 12ea

파트 아 사블레(41p)

레몬 버베나 비스퀴

달걀전란 44g
달걀노른자 88g
레몬 버베나 4g
레몬제스트 레몬 1/2개 분량
아몬드파우더 70g
박력분 9g
버터 27g
달걀흰자 105g
설탕 105g

유자 크림

생크림 220g
젤라틴매스 20.6g
유자초콜릿 171g
(☙ YUZU INSPIRATION)
유자주스 88g

유자 시럽

30보메 시럽(물 1000g,
설탕 1350g) 35g
물 35g
유자주스 22g

유자 & 레몬 버베나 젤

물 52g
유자주스 151g
설탕 41g
아가아 3.7g
레몬 버베나 2.5g

글레이즈

유자주스 371g
설탕 45g
NH펙틴 10g
물엿 74g
치자가루 적당량

유자 휩 크림

생크림A 150g
유자주스 113g
젤라틴매스 27g
화이트초콜릿 136g
(☙ IVOIRE 35%)
생크림B 347g

장식물

코코넛 분말
레몬 버베나
식용금박

기타

데코스노우

PATE A SABLÉ(41p)

LEMON VERBENA BISCUIT

44g Whole eggs
88g Egg yolks
4g Lemon verbena
Zest of 1/2 lemon
70g Almond powder
9g Cake flour
27g Butter
105g Egg whites
105g Sugar

YUJA CREAM

220g Fresh cream
20.6g Gelatin mass
171g Yuja chocolate
(☙ YUZU INSPIRATION)
88g Yuja juice

YUJA SYRUP

35g 30 Baume syrup
(1000g Water,
1350g Sugar)
35g Water
22g Yuja juice

YUJA & LEMON VERBENA GEL

52g Water
151g Yuja juice
41g Sugar
3.7g Agar-agar
2.5g Lemon verbena

GLAZE

371g Yuja juice
45g Sugar
10g Pectin NH
74g Corn syrup
QS Gardenia powder

YUJA WHIPPED CREAM

150g Fresh cream A
113g Yuja juice
27g Gelatin mass
136g White chocolate
(☙ IVOIRE 35%)
347g Fresh cream B

DECORATION

Coconut powder
Lemon verbena leaves
Edible gold leaves

OTHER

Decosnow

LEMON VERBENA BISCUIT

Process

**레몬 버베나
비스퀴**

1. 볼에 달걀흰자를 넣고 휘핑한다.

2. 설탕을 3~4회 나누어 넣어가며 계속해서 휘핑한다.

3. 거품기로 떠 올렸을 때 머랭 뿔이 탄력이 있으면서도 부드럽게 휘어지는 상태가 될 때까지 휘핑한다.

4. 달걀전란, 달걀노른자, 레몬제스트, 레몬 버베나를 핸드블렌더로 믹싱한다.

5. 볼에 옮겨준 후 아몬드파우더를 혼합한다.

6. 박력분과 55~60℃로 녹인 버터를 순서대로 넣고 혼합한다.

7. 머랭을 2회 나누어 혼합한다.

8. 32.5×32.5cm, 높이 1cm 실리콘 몰드에 완성한 레몬 버베나 비스퀴 반죽을 팬닝한 후
   160℃ 오븐에서 약 15분간 굽는다.

9. 비스퀴가 완전히 식으면 껍질이 달라붙지 않도록 비스퀴 표면에 데코스노우를 소량 뿌려준 후
   뒤집어 몰드를 제거한다.

10. 지름 7cm 원형 커터로 커팅한다.

**LEMON VERBENA
BISCUIT**

1. Whip egg whites in a mixing bowl.

2. Continue to whip, gradually adding sugar in 3~4 times.

3. Whip until the meringue forms elastic peaks and bends slightly when the whisk is
   lifted.

4. Mix whole eggs, egg yolks, lemon zest, and lemon verbena using a hand blender.

5. Transfer to a bowl and mix in almond powder.

6. Mix with cake flour and then butter melted to 55~60℃ in order.

7. Fold in the meringue in 2 times.

8. Pour finished lemon verbena biscuit batter evenly into 32.5 × 32.5 × 1cm(height)
   silicone mold. Bake for 15 minutes in the oven preheated to 160℃.

9. After the biscuit is completely cooled, sprinkle a little bit of decosnow on the surface to
   prevent the skin from sticking, then turn it upside down to remove the mold.

10. Cut using a 7cm diameter round cutter.

YUJA CREAM

11

12

13

YUJA & LEMON VERBENA GEL

14

15

16-1

16-2

유자 크림

11. 생크림을 45℃로 데운 후 녹인 젤라틴매스를 혼합한다.

12. 40℃로 녹인 유자초콜릿에 **11**을 조금씩 넣어가며 섞어준 다음 핸드블렌더로 혼합한다.

13. 30℃로 데운 유자주스를 넣고 계속해서 핸드블렌더로 혼합한 후 냉장고에서 12시간 휴지시켜 사용한다.

유자 & 레몬 버베나 젤

14. 냄비에 물과 유자주스를 넣고 가열한다. 45℃가 되면 미리 혼합해둔 설탕과 아가아가를 넣고 끓어오를 때까지 가열한다.

15. 볼에 옮겨준 후 냉장고에서 굳힌다.

16. 레몬 버베나 잎을 넣고 핸드블렌더를 이용해 부드러운 젤 상태가 될 때까지 갈아준다.

YUJA CREAM

11. Heat fresh cream to 45℃, then stir in melted gelatin mass.

12. Gradually stir in Yuja chocolate melted to 40℃ into **11**, little by little. Mix using a hand blender.

13. Add yuja juice warmed to 30℃ and continue to mix with a hand blender. Let it rest in the fridge for 12 hours before use.

YUJA & LEMON VERBENA GEL

14. In a saucepan, heat water and yuja juice. When the temperature reaches 45℃, stir in previously mixed sugar and agar-agar mixture and heat until it boils.

15. Remove to a bowl and store in the fridge to set.

16. Add lemon verbena leaves and mix using a hand blender until the mixture turns into a soft gel.

YUJA WHIPPED CREAM

17

18

19

20 YUJA SYRUP

21-1

21-2

유자 휩 크림	17.	끓기 직전까지 가열한 생크림A에 젤라틴매스를 넣고 혼합한다.
	18.	35℃로 녹인 화이트초콜릿에 **17**을 붓고 핸드블렌더로 혼합한다.
	19.	35℃로 데운 생크림B와 유자주스를 순서대로 넣고 계속해서 핸드블렌더로 혼합한다.
	20.	냉장고에서 12시간 휴지시킨 후 휘핑해 사용한다.
유자 시럽	21.	30보메 시럽, 물, 유자주스를 혼합한다.

YUJA WHIPPED CREAM	17.	Stir gelatin mass into fresh cream A heated until just before boiling.
	18.	Pour **17** over white chocolate melted to 35℃. Mix using a hand blender.
	19.	Add fresh cream B warmed to 35℃, then yuja juice in order. Continue to mix using a hand blender.
	20.	Rest in the fridge for 12 hours. Whip to use.
YUJA SYRUP	21.	Mix 30 Baume syrup, water, and yuja juice.

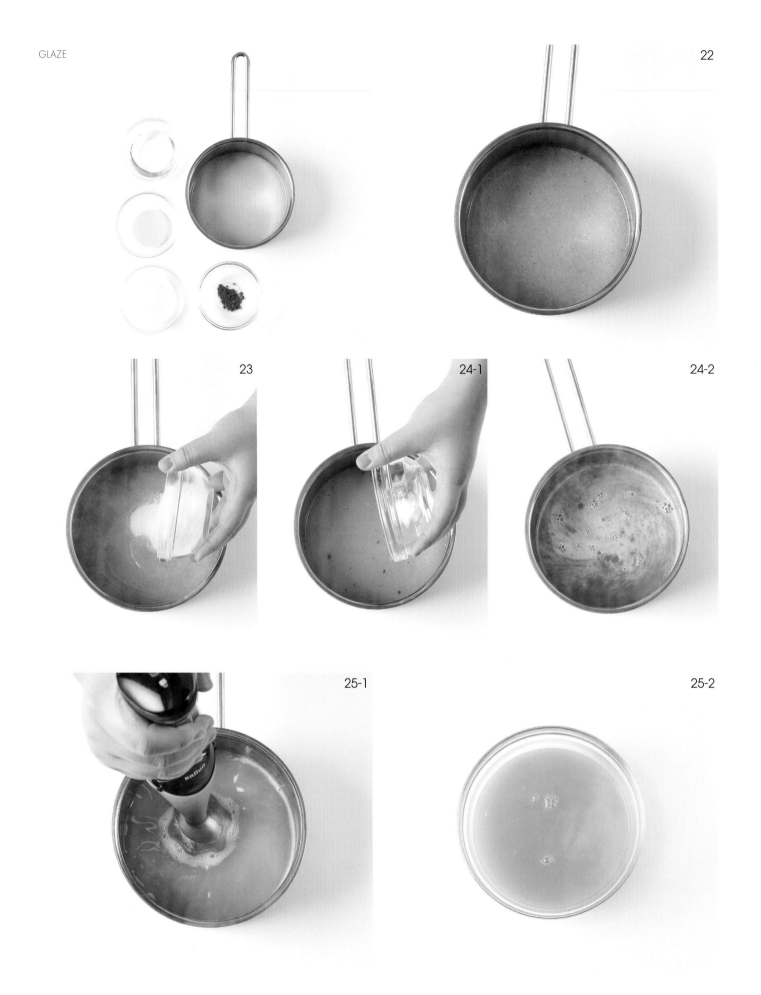

글레이즈	22.	냄비에 유자주스와 소량의 치자가루를 넣고 45℃로 데워준다.
	23.	미리 혼합해둔 NH펙틴과 설탕을 넣고 혼합한다.
	24.	물엿을 넣고 끓어오를 때까지 가열한다.
	25.	사용할 때는 65℃로 맞춰준 후 핸드블렌더로 균일하게 믹싱해 사용한다.

GLAZE	22.	In a saucepan, heat yuja juice and a pinch of gardenia powder to 45℃.
	23.	Stir in previously mixed pectin NH and sugar mixture.
	24.	Add corn syrup and heat until the mixture boils.
	25.	To use, set the glaze to 65℃, and mix thoroughly using a hand blender.

ASSEMBLY

26

27

28

29-1

29-2

30

FINISH 31

32

33

34

35

조립	26.	레몬 버베나 비스퀴에 유자 시럽을 적셔준다.
	27.	100% 소성한 타르트 셸에 유자&레몬 버베나 젤을 15g씩 파이핑한다.
	28.	시럽에 적신 레몬 버베나 비스퀴를 유자&레몬 버베나 젤 위에 얹고 평평한 도구로 가볍게 눌러 평탄 작업을 한다.
	29.	유자 크림을 가득 채운 후 스패츌러를 이용해 표면을 고르게 정리한다. 냉동고에서 단단하게 굳힌다.
	30.	남은 유자 크림은 지름 2cm 반구형 실리콘 몰드에 채워 냉동한다.

마무리	31.	생토노레(15번) 깍지를 이용해 유자 휩 크림을 모양내어 파이핑한 후 냉동고에서 단단하게 굳힌다.
	32.	유자 휩 크림 표면에 크리스탈 스프레이 건 믹스처(73p)를 분사한다.
	33.	몰드에서 빼낸 반구 형태의 유자 크림을 글레이징한다.
	34.	가장자리에 코코넛 분말을 묻혀준 후 타르트 중앙에 올린다.
	35.	레몬 버베나와 식용금박으로 장식해 마무리한다.

ASSEMBLY	26.	Brush yuja syrup on lemon verbena biscuits.
	27.	In each 100% baked tarte shells, pipe 15g of yuja & lemon verbena gel.
	28.	Place lemon verbena biscuits soaked with syrup on top of yuja & lemon verbena gel, and lightly press with a flat tool.
	29.	Fill with yuja cream completely, and even out the surface using a spatula. Freeze completely.
	30.	Fill remaining yuja cream in 2cm diameter half-sphere silicone molds and store in the freezer.

FINISH	31.	Pipe yuja whipped cream using a St. Honore nozzle (#15) and freeze completely.
	32.	Spray crystal spray gun mixture (73p) on the surface of the frozen yuja whipped cream.
	33.	Remove the frozen half-sphere yuja cream from the molds. Glaze the frozen cream.
	34.	Coat the bottom edge with coconut powder and place it on the center of the tarte.
	35.	Decorate with lemon verbena leaves and edible gold leaves to finish.

RASPBERRY BLOSSOM
TARTE

라즈베리 블로섬 타르트

ingredients - ∅ 16cm 2ea

파트 아 사블레(41p)
or **스위트 반죽 – 글루
텐 프리**(53p)

GLUTEN
FREE

장식물

코코넛 분말

스위트 바질

식용금박

라즈베리 크림

생크림 224g

젤라틴매스 19.8g

프람보아즈초콜릿 149g

(⚜ FRAMBOISE INSPIRATION)

물 63g

라즈베리 퓌레 111g

라임 & 바질 젤

물 53g

라임주스 149g

설탕 41g

아가아가 3.7g

스위트 바질 3.5g

바질 휩 크림

우유 81g

젤라틴매스 12g

화이트초콜릿 72g

(⚜ IVOIRE 35%)

생크림 246g

스위트 바질 6g

라즈베리주스＊

냉동 라즈베리 500g

설탕 76g

글레이즈

라즈베리주스＊ 252g

NH펙틴 3.5g

설탕 10g

물엿 25g

PATE A SABLÉ(41p)
or **SWEET DOUGH -
GLUTEN FREE**(53p)

GLUTEN
FREE

DECORATION

Coconut powder

Sweet basil

Edible gold leaves

RASPBERRY CREAM

224g Fresh cream

19.8g Gelatin mass

149g Framboise
chocolate
(⚜ FRAMBOISE
INSPIRATION)

63g Water

111g Raspberry purée

LIME & BASIL GEL

53g Water

149g Lime juice

41g Sugar

3.7g Agar-agar

3.5g Sweet basil

**BASIL WHIPPED
CREAM**

81g Milk

12g Gelatin mass

72g White chocolate
(⚜ IVOIRE 35%)

246g Fresh cream

6g Sweet basil

RASPBERRY JUICE＊

500g Frozen
raspberries

76g Sugar

GLAZE

252g Raspberry juice＊

3.5g Pectin NH

10g Sugar

25g Corn syrup

RASPBERRY CREAM

1

2

3

4

LIME & BASIL GEL

5

6-1

6-2

Process

라즈베리 크림

1. 생크림을 45℃로 데운 후 녹인 젤라틴매스를 넣고 혼합한다.

2. 40℃로 녹인 프람보아즈초콜릿에 **1**을 조금씩 나눠 넣어가며 섞어준 다음 핸드블렌더로 혼합한다.

3. 라즈베리 퓌레와 물을 30℃로 데운 후 **2**에 부어준다.

4. 계속해서 핸드블렌더로 혼합한 후 냉장고에서 12시간 휴지시켜 사용한다.

라임 & 바질 젤

5. 냄비에 물과 라임주스를 넣고 가열한다. 45℃가 되면 미리 혼합해둔 설탕과 아가아가를 넣고
끓어오를 때까지 가열한 후 볼에 옮겨 냉장고에서 굳힌다.

6. 스위트 바질을 넣고 푸드프로세서를 이용해 부드러운 젤 상태가 될 때까지 갈아준다.

**RASPBERRY
CREAM**

1. Heat fresh cream to 45℃, then stir in melted gelatin mass.

2. Gradually stir **1** in Framboise chocolate melted to 40℃, little by little, then mix using
a hand blender.

3. Warm raspberry purée and water to 30℃. Pour into **2**.

4. Continue to mix with a hand blender. Let it rest in the fridge for 12 hours before use.

**LIME & BASIL
GEL**

5. In a saucepan, heat water and lime juice. When the temperature reaches 45℃,
stir in previously mixed sugar and agar-agar mixture and heat until it boils.
Remove to a bowl and store in the fridge to set.

6. Add sweet basil and mix using a hand blender until the mixture turns into
a soft gel.

BASIL WHIPPED CREAM

7

8

9

10   RASPBERRY JUICE

11

12

13

바질 휩 크림

7. 우유를 끓기 직전까지 가열한 후 젤라틴매스를 혼합한다.

8. 35℃로 녹인 화이트초콜릿에 **7**을 붓고 핸드블렌더로 혼합한다.

9. 35℃로 데운 생크림과 스위트 바질을 넣고 계속해서 핸드블렌더로 혼합한 후 체에 걸러준다.

10. 냉장고에서 12시간 휴지시킨 후 휘핑해 사용한다.

라즈베리주스

11. 냉동 라즈베리와 설탕을 수비드 백에 담는다.

12. 약 60℃로 1시간 동안 수비드한다.

13. 체에 걸러 라즈베리주스를 추출한다.

BASIL WHIPPED CREAM

7. Stir gelatin mass into milk heated until just before boiling.

8. Pour **7** over white chocolate melted to 35℃. Mix using a hand blender.

9. Add fresh cream warmed to 35℃ and sweet basil leaves. Continue to mix using a hand blender, then strain.

10. Rest in the fridge for 12 hours. Whip to use.

RASPBERRY JUICE

11. Pour frozen raspberries and sugar into a sous vide bag.

12. Sous vide at 60℃ for 1 hour.

13. Strain to extract raspberry juice.

14

15

16

17-1

17-2

8

글레이즈	14.	NH펙틴과 설탕을 혼합한다.
	15.	라즈베리주스의 무게(252g)를 확인한 후 45℃까지 가열한다.
	16.	**14**를 넣고 섞는다.
	17.	물엿을 넣고 끓어오를 때까지 가열한다.
	18.	사용할 때는 65℃로 맞춰준 후 핸드블렌더로 균일하게 믹싱해 사용한다.

GLAZE	14.	Mix pectin NH and sugar.
	15.	Check the weight of the raspberry juice (252g). Heat until 45℃.
	16.	Stir in **14**.
	17.	Stir in corn syrup and heat until it boils.
	18.	To use, set the glaze to 65℃, and mix thoroughly using a hand blender.

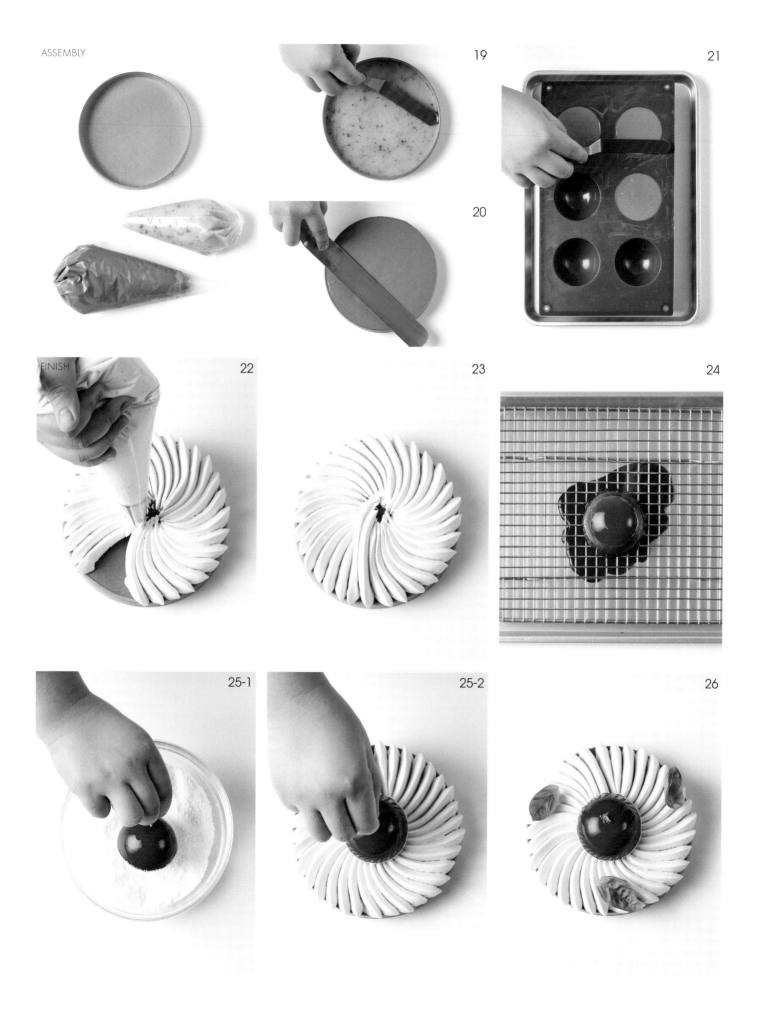

19

20

21

22

23

24

25-1

25-2

26

조립	19.	100% 소성한 타르트 셸에 라임&바질 젤을 90g씩 파이핑한 후 평탄 작업을 한다.
	20.	라임&바질 젤 위에 라즈베리 크림을 가득 채운 후 스패출러를 이용해 표면을 고르게 정리한다. 냉동고에서 단단하게 굳힌다.
	21.	남은 라즈베리 크림은 지름 6cm 반구형 실리콘 몰드에 채워 냉동한다.
마무리	22.	생토노레(15번) 깍지를 이용해 바질 휩 크림을 모양내어 파이핑한 다음 냉동고에서 단단하게 굳힌다.
	23.	바질 휩 크림 표면에 크리스탈 스프레이 건 믹스처를 분사한다.
	24.	몰드에서 빼낸 반구 형태의 라즈베리 크림을 글레이징한다.
	25.	가장자리에 코코넛 분말을 묻혀준 후 타르트 중앙에 올린다.
	26.	스위트 바질과 식용금박으로 장식해 마무리한다.

ASSEMBLY	19.	In each 100% baked tarte shells, pipe 90g of lime & basil gel. Spread evenly.
	20.	Fill with raspberry cream on top of the lime & basil gel completely, and even out the surface using a spatula. Freeze completely.
	21.	Fill remaining raspberry cream in 6cm diameter half-sphere silicone molds and store in the freezer.
FINISH	22.	Pipe basil whipped cream using a St. Honore nozzle (#15) and freeze completely.
	23.	Spray crystal spray gun mixture on the surface of the basil whipped cream.
	24.	Remove the frozen half-sphere raspberry cream from the molds. Glaze the frozen cream.
	25.	Coat the bottom edge with coconut powder and place it on the center of the tarte.
	26.	Decorate with sweet basil leaves and edible gold leaves to finish.

³ APRICOT TARTE

살구 타르트

ingredients · ⌀ 7cm 15ea

파트 아 사블레(41p)
or 스위트 반죽 – 글루텐 프리(53p)

GLUTEN
FREE

살구 충전물

살구 220g
올리브오일 적당량
살구 & 복숭아 젤* 176g
코리엔더 1g
타라곤 2g

타라곤 풍미의 라이트 아몬드 크림

아몬드파우더 90g
슈거파우더 70g
달걀전란 80g
타라곤 2g
전분 13g
버터 15g
달걀흰자 24g
설탕 24g

살구 & 복숭아 젤*

복숭아 퓌레 53g
살구 퓌레 124g
레몬주스 27g
설탕 41g
아가아가 3.7g

살구 & 마스카르포네 휩 크림

살구 퓌레 65g
젤라틴매스 7.2g
화이트초콜릿 72g
(IVOIRE 35%)
생크림 183g
마스카르포네치즈 78g

장식물

허브

글레이즈

물 72g
살구 퓌레 168g
NH펙틴 4.5g
설탕 20g
물엿 34g
오렌지색 천연 식용 색소 적당량

기타

살구

PATE A SABLÉ(41p)
or SWEET DOUGH - GLUTEN FREE(53p)

GLUTEN
FREE

APRICOT FILLING

220g Apricots
QS Olive oil
176g Apricot & peach gel*
1g Coriander seeds
2g Tarragon leaves

LIGHT ALMOND CREAM WITH TARRAGON AROMA

90g Almond powder
70g Powdered sugar
80g Whole eggs
2g Tarragon leaves
13g Starch
15g Butter
24g Egg whites
24g Sugar

APRICOT & PEACH GEL*

53g Peach purée
124g Apricot purée
27g Lemon juice
41g Sugar
3.7g Agar-agar

APRICOT & MASCARPONE WHIPPED CREAM

65g Apricot purée
7.2g Gelatin mass
72g White chocolate
(IVOIRE 35%)
183g Fresh cream
78g Mascarpone cheese

DECORATION

Herbs

GLAZE

72g Water
168g Apricot purée
4.5g Pectin NH
20g Sugar
34g Corn syrup
QS Orange natural food color

OTHER

Apricots

LIGHT ALMOND CREAM
WITH TARRAGON AROMA

Process

**타라곤 풍미의
라이트
아몬드 크림**

1. 볼에 달걀흰자를 넣고 휘핑한다.

2. 설탕을 3~4회 나누어 넣어가며 계속해서 휘핑한다.

3. 거품기로 떠 올렸을 때 머랭 뿔이 탄력이 있으면서도 부드럽게 휘어지는 상태가 될 때까지 휘핑한다.

4. 볼에 달걀전란과 타라곤을 넣고 핸드블렌더로 믹싱한다.

5. 아몬드파우더와 슈거파우더를 넣고 섞어준 후 휘핑한다.

6. 체 친 전분과 55~60℃로 녹인 버터를 순서대로 혼합한다.

7. 머랭을 2회 나누어 혼합한다.

8. 80% 소성한 타르트 셸에 완성한 타라곤 풍미의 라이트 아몬드 크림을 15g씩 채워준다.

9. 세그먼트한 살구를 적당량 얹고 160℃ 오븐에서 15분간 굽는다.

**LIGHT ALMOND
CREAM WITH
TARRAGON
AROMA**

1. Whip egg whites in a mixing bowl.

2. Continue to whip, gradually adding sugar in 3~4 times.

3. Whip until the meringue forms elastic peaks and bends slightly when the whisk is lifted.

4. In a bowl, mix whole eggs and tarragon leaves using a hand blender.

5. Whisk in almond powder and powdered sugar, and whip.

6. Mix with sifted starch and then butter melted to 55~60℃ in order.

7. Fold in the meringue in 2 times.

8. Pipe 15g of light almond cream with tarragon aroma in 80% baked tarte shells.

9. Top with apricot segments and bake for 15 minutes at 160℃.

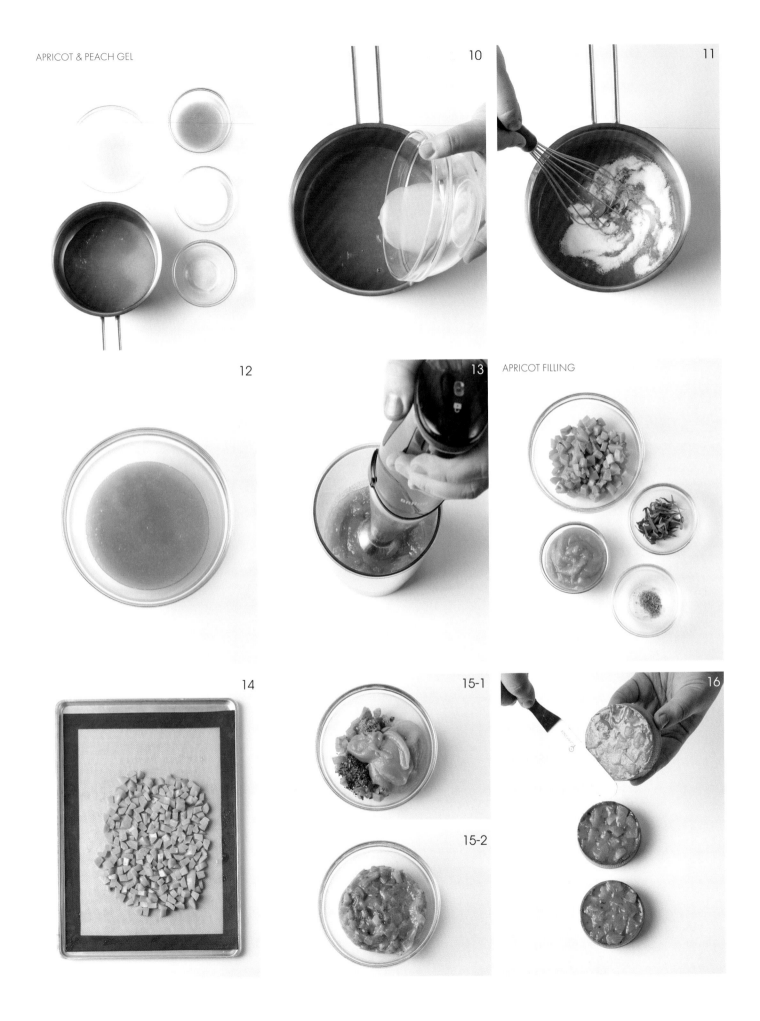

APRICOT & PEACH GEL

10

11

12

13

APRICOT FILLING

14

15-1

15-2

16

살구 & 복숭아 젤	10.	냄비에 복숭아 퓌레, 살구 퓌레, 레몬주스를 넣고 45℃로 가열한다.
	11.	충분히 혼합한 설탕과 아가아가를 넣고 끓어오를 때까지 가열한다.
	12.	냉장고에서 굳힌다.
	13.	핸드블렌더를 이용해 부드러운 젤 상태가 될 때까지 갈아준다.
살구 충전물	14.	사방 1cm 크기로 자른 살구에 올리브유를 적당량 넣고 버무린 후 250℃ 오븐에서 약 2분간 구워준다.
	15.	완전히 식힌 살구, 잘게 다진 타라곤, 분쇄한 코리엔더 씨드, 살구&복숭아 젤을 혼합한다.
	16.	타라곤 풍미의 라이트 아몬드 크림 위에 완성한 살구 충전물을 가득 채워준다.

APRICOT & PEACH GEL	10.	In a saucepan, heat peach purée, apricot purée, and lemon juice to 45℃.
	11.	Stir in previously thoroughly mixed sugar and agar-agar and heat until it boils.
	12.	Store in the fridge to set.
	13.	Blend with a hand blender until it turns into a soft gel.
APRICOT FILLING	14.	Toss apricots diced in 1cm cubes in a moderate amount of olive oil and bake for about 2 minutes at 250℃.
	15.	Mix completely cooled apricots, finely chopped tarragon leaves, ground coriander seeds and apricot & peach gel.
	16.	Top it off with finished apricot filling on light almond cream with tarragon aroma. Freeze completely.

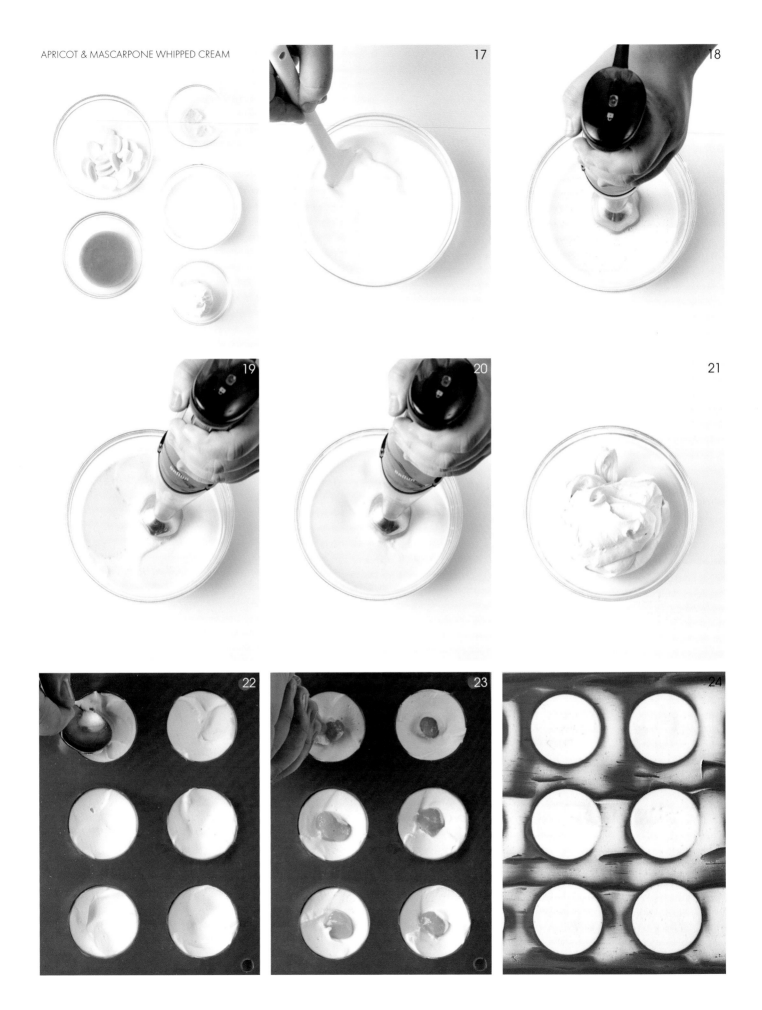

살구 & 마스카르포네
휩 크림

17. 40℃로 데운 생크림과 녹인 젤라틴매스를 혼합한다.

18. 35℃로 녹인 화이트초콜릿에 **17**을 붓고 핸드블렌더로 혼합한다.

19. 35℃로 데운 살구 퓌레를 넣고 혼합한다.

20. 마스카르포네치즈를 넣고 계속해서 혼합한다.

21. 냉장고에서 12시간 휴지시킨 후 휘핑해 사용한다.

22. 지름 4cm 반구형 실리콘 몰드에 살구&마스카르포네 휩 크림을 50% 정도 채운 후 가운데가 오목하도록 작업한다.

23. 몰드 정중앙에 완성한 살구&복숭아 젤을 약 3g 파이핑한다.

24. 다시 살구&마스카르포네 휩 크림을 가득 채운다. 스패츌러로 표면을 고르게 정리한 후 냉동고에서 단단하게 굳힌다.

APRICOT &
MASCARPONE
WHIPPED CREAM

17. Mix fresh cream heated to 40℃ and melted gelatin.

18. Pour **17** over white chocolate melted to 35℃, and mix with a hand blender.

19. Mix with apricot purée warmed to 35℃.

20. Add mascarpone cheese and continue mixing.

21. Rest in the fridge for 12 hours. Whip to use.

22. Fill 50% of 4cm diameter half-sphere silicone molds with apricot & mascarpone whipped cream. Concave the center with a spoon.

23. Pipe about 3g of finished apricot & peach gel in the very center of the molds.

24. Top it off with apricot & mascarpone whipped cream again. Even out the surface with a spatula and store in the freezer until completely frozen.

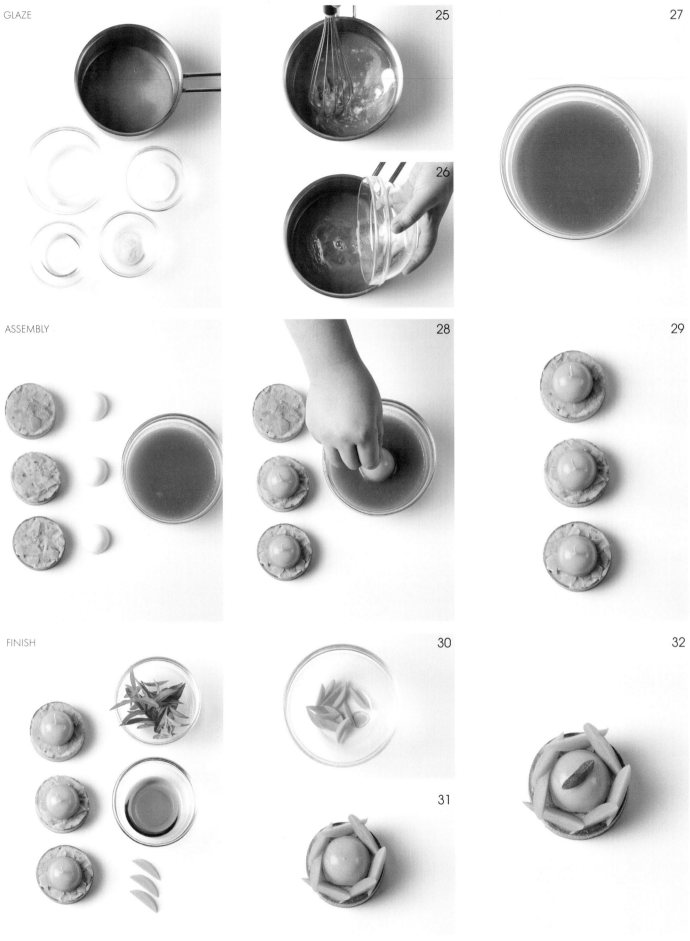

GLAZE

25

26

27

ASSEMBLY

28

29

FINISH

30

31

32

글레이즈	25.	냄비에 물, 살구 퓌레, 오렌지색 천연 식용 색소를 넣고 45℃로 가열한다.
	26.	미리 혼합해둔 NH펙틴과 설탕을 넣고 혼합한다. 물엿을 넣고 끓어오를 때까지 가열한다.
	27.	사용할 때는 65℃로 맞춰준 후 핸드블렌더로 균일하게 믹싱해 사용한다.
조립	28.	실리콘 몰드에서 빼낸 살구&마스카포네 휩 크림을 글레이징한다.
	29.	타르트 중앙에 올린다.
마무리	30.	세그먼트한 살구에 소량의 올리브유를 넣고 버무린다.
	31.	타르트 가장자리에 둘러준다.
	32.	허브로 장식해 마무리한다.

GLAZE	25.	In a saucepan, heat water, apricot purée, and orange natural food color until 45℃.
	26.	Stir in previously mixed pectin NH and sugar. Add corn syrup and continue to heat until it boils.
	27.	To use, set the glaze to 65℃, and mix thoroughly using a hand blender.
ASSEMBLY	28.	Remove the frozen apricot & mascarpone whipped cream from the molds, and glaze the cream.
	29.	Place on the center of the tartes.
FINISH	30.	Coat segmented apricots with a little bit of olive oil.
	31.	Place around the edge of the tartes.
	32.	Decorate with herbs to finish.

⁴ RASPBERRY & PASSION FRUIT TARTE

산딸기 & 패션후르츠 타르트

ingredients - Ø 8cm 15ea

파트 아 사블레(41p) or
스위트 반죽 – 글루텐 프리
(53p)

GLUTEN
FREE

**라임과 민트 풍미의
라이트 아몬드 크림**

아몬드파우더 90g
슈거파우더 70g
달걀전란 80g
애플민트 3g
라임제스트 라임 1/3개 분량
전분 13g
버터 15g
달걀흰자 24g
설탕 24g

패션후르츠 크림

달걀전란 144g
설탕 101g
레몬주스 13g
패션후르츠 퓌레 103g
젤라틴매스 10.5g
버터 200g

장식물

데코젤
허브
식용꽃

라즈베리 잼

냉동 라즈베리 145g
라즈베리 퓌레 145g
설탕 80g
NH펙틴 4g
레몬주스 13g
키르쉬 13g

기타

산딸기

PATE A SABLÉ(41p) or **SWEET
DOUGH - GLUTEN FREE**(53p)

GLUTEN
FREE

**LIGHT ALMOND CREAM
WITH LIME & MINT AROMA**

90g Almond powder
70g Powdered sugar
80g Whole eggs
3g Applemint
Zest of 1/3 lime
13g Starch
15g Butter
24g Egg whites
24g Sugar

PASSION FRUITS CREAM

144g Whole eggs
101g Sugar
13g Lemon juice
103g Passion fruits purée
10.5g Gelatin mass
200g Butter

DECORATION

Decogel
Herbs
Edible flowers

RASPBERRY JAM

145g Frozen raspberries
145g Raspberry purée
80g Sugar
4g Pectin NH
13g Lemon juice
13g Kirsch

OTHER

Raspberries

LIGHT ALMOND CREAM WITH LIME & MINT
AROMA

Process

라임과 민트 풍미의
라이트 아몬드
크림

1. 볼에 달걀흰자를 넣고 휘핑한다.

2. 설탕을 3~4회 나누어 넣어가며 계속해서 휘핑한다.

3. 거품기로 떠 올렸을 때 머랭 뿔이 탄력이 있으면서도 부드럽게 휘어지는 상태가 될 때까지 휘핑한다.

4. 볼에 달걀전란, 애플민트, 라임제스트를 넣고 핸드블렌더로 믹싱한다.

5. 아몬드파우더와 슈거파우더를 넣고 섞어준 후 휘핑한다.

6. 체 친 전분과 55~60℃로 녹인 버터를 순서대로 혼합한다.

7. 머랭을 2회 나누어 혼합한다.

8. 80% 소성한 타르트 셸에 라임과 민트 풍미의 라이트 아몬드 크림을 15g씩 채워준다.

9. 산딸기를 적당량 얹고 160℃ 오븐에서 15분간 굽는다.

LIGHT ALMOND
CREAM WITH
LIME & MINT
AROMA

1. Whip egg whites in a mixing bowl.

2. Continue to whip, gradually adding sugar in 3~4 times.

3. Whip until the meringue forms elastic peaks and bends slightly when the whisk is lifted.

4. In a bowl, mix whole eggs, applemint, and lime zest using a hand blender.

5. Whisk in almond powder and powdered sugar, and whip.

6. Mix with sifted starch and then butter melted to 55~60℃ in order.

7. Fold in the meringue in 2 times.

8. Pipe 15g of light almond cream with lime & mint aroma in 80% baked tarte shells.

9. Top with a moderate amount of raspberries. Bake for 15 minutes at 160℃.

패션후르츠 크림

10. 볼에 달걀전란과 설탕 1/2을 넣고 혼합한다.

11. 냄비에 패션후르츠 퓌레, 레몬주스, 나머지 설탕을 넣고 45℃로 가열한다.

12. 10에 11을 조금씩 부어가며 섞어준다.

13. 12를 다시 냄비에 옮겨 75~78℃까지 가열한다.

14. 젤라틴매스를 넣고 혼합한다.

15. 체에 걸러준 후 45℃까지 식힌다.

16. 상온 상태의 버터를 넣고 핸드블렌더로 혼합한다.

17. 라임과 민트 풍미의 라이트 아몬드 크림 위에 완성한 패션후르츠 크림을 35g씩 채운다.

18. 냉동고에 넣어 단단하게 굳힌다.

PASSION FRUITS CREAM

10. In a bowl, mix whole eggs and 1/2 of sugar.

11. In a saucepan, heat passion fruits purée, lemon juice and remaining sugar until 45℃.

12. Gradually mix 11 into 10, a little bit at a time.

13. Pour 12 back into a saucepan and heat until 75~78℃.

14. Stir in gelatin mass.

15. Sieve and cool it down to 45℃.

16. Incorporate room temperature butter using a hand blender.

17. Fill 35g of finished passion fruits cream on top of the light almond cream with lime & mint aroma

18. Freeze completely.

RASPBERRY JAM

19

20-1

20-2

21

22

23

FINISH

24

25

라즈베리 잼　　19.　냄비에 냉동 라즈베리와 라즈베리 퓌레를 넣고 45℃로 가열한다.

　　20.　미리 혼합한 설탕과 NH펙틴을 넣고 끓어오를 때까지 가열한다.

　　21.　불에서 내린 후 레몬주스와 키르쉬를 순서대로 넣고 섞어준다.

　　22.　완성된 잼은 차갑게 식힌 후 사용한다.

　　23.　냉동고에서 단단하게 굳힌 패션후르츠 크림 위에 라즈베리 잼을 채운다.

마무리　　24.　신선한 제철 산딸기를 가득 올린 후 라즈베리 퓌레를 섞은 데코젤을 산딸기에 채워준다.

　　25.　허브, 식용꽃으로 장식해 마무리한다.

RASPBERRY JAM　　19.　In a saucepan, heat frozen raspberries and raspberry purée until 45℃.

　　20.　Stir in previously mixed sugar and pectin NH mixture, and continue to heat until the mixture boils.

　　21.　Remove from heat, and stir in lemon juice, then Kirsch in order.

　　22.　Cool finished jam completely before use.

　　23.　Fill with raspberry jam on top of the frozen passion fruits cream.

FINISH　　24.　Top with fresh seasonal raspberries, then fill the berries with decogel mixed with raspberry purée.

　　25.　Decorate with herbs and edible flowers to finish.

⁵ TROPICAL TARTE

트로피칼 타르트

ingredients - Ø 8.5cm 12ea

코코넛 스트로이젤

버터 106g
설탕 106g
소금 1.4g
아몬드파우더 66g
코코넛 분말 66g
박력분 106g

코코넛 비스퀴

달걀전란 68g
달걀노른자 26g
설탕A 84g
코코넛 분말 115g
박력분 26g
버터 58g
달걀흰자 42g
설탕B 42g

패션후르츠 크림(125p)

달걀전란 167g
설탕 116g
레몬주스 16g
패션후르츠 퓌레 119g
젤라틴매스 12g
버터 230g

로열 아이싱*

슈거파우더 50g
달걀흰자 10g
노란색 천연 식용 색소

코코넛 휩 크림

코코넛밀크 129g
우유 103g
젤라틴매스 19.8g
화이트초콜릿 116g
(IVOIRE 35%)
생크림 298g

슈거 페이스트*

슈거파우더 200g
젤라틴매스 24g
레몬주스 5g
로열아이싱* 적당량

장식물

화이트초콜릿
(OPALYS 33%)
노란색 천연 식용 색소
피스타치오
슈거 페이스트*
코코넛
식용꽃

COCONUT STREUSEL

106g Butter
106g Sugar
1.4g Salt
66g Almond powder
66g Coconut powder
106g Cake flour

COCONUT BISCUIT

68g Whole eggs
26g Egg yolks
84g Sugar A
115g Coconut powder
26g Cake flour
58g Butter
42g Egg white
42g Sugar B

PASSION FRUITS CREAM(125p)

167g Whole eggs
116g Sugar
16g Lemon juice
119g Passion fruits purée
12g Gelatin mass
230g Butter

ROYAL ICING*

50g Powdered sugar
10g Egg whites
QS Yellow natural food color

COCONUT WHIPPED CREAM

129g Coconut milk
103g Milk
19.8g Gelatin mass
116g White chocolate
(IVOIRE 35%)
298g Fresh cream

SUGAR PASTE*

200g Powdered sugar
24g Gelatin mass
5g Lemon juice
QS Royal icing*

DECORATION

White chocolate
(OPALYS 33%)
Yellow natural food color
Pistachios
Sugar paste*
Coconut
Edible flowers

1-1

1-2

2

3-1

3-2

4

5

6

Process

**코코넛
스트로이젤**

1. 버터를 부드럽게 풀어준 후 설탕과 소금을 넣고 혼합한다.

2. 아몬드파우더, 코코넛 분말, 박력분을 넣고 한 덩어리가 될 때까지 믹싱한다.

3. 두 장의 투명 필름 사이에 반죽을 놓고 2mm 두께로 밀어 편다. 이때 2mm 두께의 바를 이용하면 일정한 두께로 밀어 펼 수 있다.

4. 냉동고에 넣어 단단하게 굳힌다.

5. 지름 8.5cm 원형 틀을 이용해 커팅한다.

6. 틀 째 오븐에 넣고 160℃에서 12분간 굽는다.

**COCONUT
STREUSEL**

1. Beat butter to soften, then mix with sugar and salt.

2. Add almond powder, coconut powder, and cake flour and mix until it forms a ball.

3. Place the dough between two sheets of clear plastic film and roll to 2mm thickness. Using 2mm thick bars will help to roll out evenly.

4. Place in the freezer to harden.

5. Cut using 8.5cm diameter round ring molds.

6. Place the streusels with the molds in a 160°C oven, and bake for 12 minutes.

코코넛 비스퀴

7. 볼에 달걀흰자를 넣고 휘핑한다.

8. 설탕 B를 3~4회 나누어 넣어가며 계속해서 휘핑한다.

9. 거품기로 떠 올렸을 때 머랭 뿔이 탄력이 있으면서도 부드럽게 휘어지는 상태가 될 때까지 휘핑한다.

10. 볼에 달걀전란, 달걀노른자, 설탕A를 넣고 휘핑한다.

11. 코코넛 분말과 체 친 박력분을 넣고 섞는다.

12. 55~60℃로 녹인 버터를 넣고 섞는다.

13. 머랭을 2회 나누어 혼합한다.

14. 코코넛 스트로이젤 위에 완성한 코코넛 비스퀴 반죽을 35g씩 파이핑한 후 160℃ 오븐에서 15분간 굽는다.

COCONUT BISCUIT

7. Whip egg whites in a mixing bowl.

8. Continue to whip, gradually adding sugar B in 3~4 times.

9. Whip until the meringue forms elastic peaks and bends slightly when the whisk is lifted.

10. In a bowl, whip whole eggs, egg yolks, and sugar A.

11. Mix with coconut powder and sifted cake flour.

12. Stir in butter melted to 55~60℃.

13. Fold in the meringue in 2 times.

14. Pipe 35g of finished coconut biscuit on top of the coconut streusels, then bake at 160℃ for 15 minutes.

COCONUT WHIPPED CREAM

15

16

17

18 ASSEMBLY

19

20

21

코코넛 휩 크림   **15.**   코코넛밀크와 우유를 끓기 직전까지 가열한 후 젤라틴매스를 넣고 혼합한다.

       16.   35℃로 녹인 화이트초코릿에 **15**을 붓고 핸드블렌더로 혼합한다.

       17.   35℃로 데운 생크림을 넣고 계속해서 핸드블렌더로 혼합한다.

       18.   냉장고에서 12시간 휴지시킨 후 휘핑해 사용한다.

조립   **19.**   크림 파이핑 작업 전 코코넛 비스퀴와 타르트 링이 잘 분리되도록 작업한다.

       20.   코코넛 비스퀴 위에 코코넛 휩 크림과 패션후르츠 크림을 번갈아가며 파이핑한다.

       * 원형(지름 1cm) 깍지를 이용해 원뿔 모양으로 파이핑한다.

       21.   냉동고에서 단단하게 굳힌다.

COCONUT WHIPPED CREAM   **15.**   Heat coconut milk and milk until just before boiling. Stir in gelatin mass.

       16.   Pour **15** over white chocolate melted to 35℃. Mix using a hand blender.

       17.   Add fresh cream warmed to 35℃, and continue to mix with a hand blender.

       18.   Rest in the fridge for 12 hours. Whip to use.

ASSEMBLY   **19.**   Before piping the cream, run a knife along the ring mold to separate the coconut biscuit.

       20.   Pipe coconut whipped cream and passion fruits cream alternately on top of the coconut biscuits.

       * Use a round tip (1cm diameter) to pipe in cone shapes.

       21.   Freeze completely.

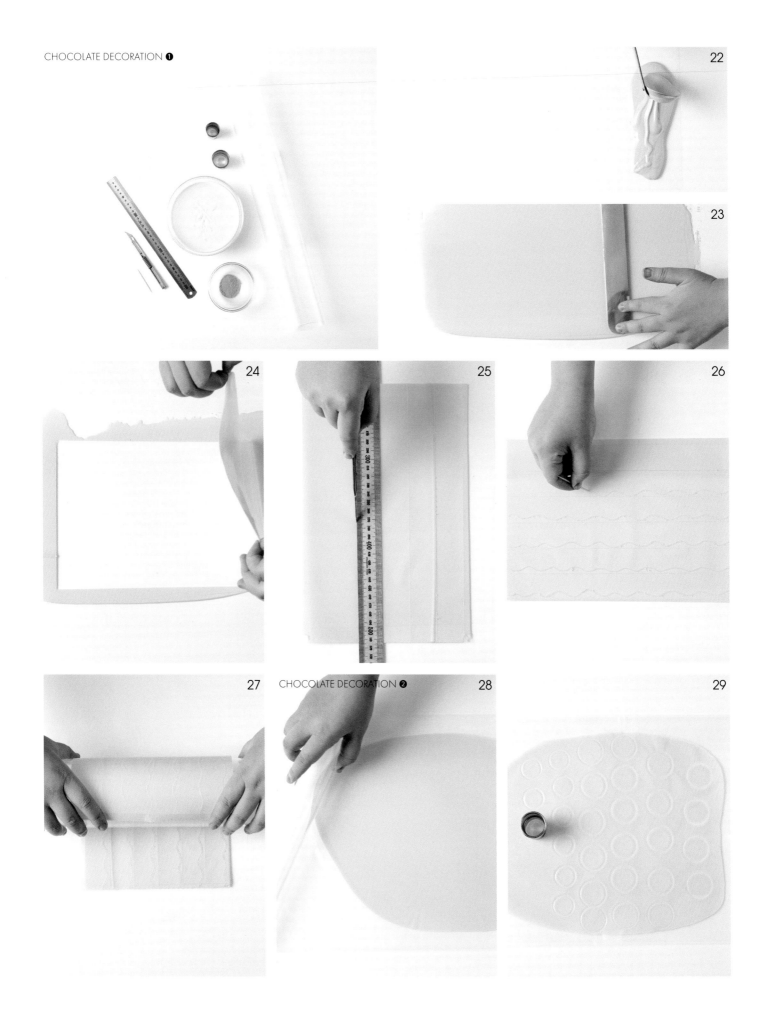

초콜릿 **장식물** **❶**	22.	투명 필름 위에 노란색 천연 식용 색소를 섞어 템퍼링한 화이트초콜릿을 적당량 부어준다.
	23.	스패출러를 이용해 얇게 밀어 편다.
	24.	초콜릿이 굳기 시작하면 조심스럽게 필름을 들어올린다.
	25.	폭 3.3cm, 길이 27cm로 커팅한다.
	26.	뾰족한 도구를 이용해 물결 모양으로 커팅한다.
	27.	지름 8cm 아크릴 파이프에 감아준다.
초콜릿 **장식물** **❷**	28.	두 장의 투명 필름 사이에 동일한 초콜릿을 적당량 붓고 밀대를 이용해 균일한 두께로 밀어 편다.
	29.	지름 2.5cm, 지름 3.5cm 원형 커터로 커팅한다.

CHOCOLATE **DECORATION** **❶**	22.	Pour a moderate amount of tempered white chocolate mixed with yellow natural food color on top of a clear plastic film.
	23.	Spread thinly using a spatula.
	24.	When the chocolate starts to crystallize, carefully lift the film.
	25.	Cut into 3.3cm wide, 27cm long.
	26.	Using a pointed tool, draw wavy lines to cut through the chocolates.
	27.	Wrap around an 8cm diameter acrylic pipe.
CHOCOLATE **DECORATION** **❷**	28.	Between two sheets of clear plastic film, pour a moderate amount of the same chocolate and roll into even thickness using a rolling pin.
	29.	Cut with 2.5cm and 3.5cm diameter round cutters.

SUGAR PASTE

30

31

32

LOYAL ICING

33

34

FINISH

35

36

37

슈거 페이스트	30.	믹싱볼에 슈거파우더, 레몬주스, 녹인 젤라틴매스를 넣고 한 덩어리가 될 때까지 믹싱한다.
	31.	완성한 반죽을 최대한 얇게 밀어 편다.
	32.	원하는 모양의 커터로 커팅한 후 상온에서 하루 동안 건조시킨다.
로열 아이싱	33.	슈거파우더, 달걀흰자, 노란색 천연 식용 색소 소량을 혼합한 후 휘핑한다.
	34.	건조한 슈거 페이스트 꽃 장식물 위에 동그랗게 파이핑한 후 상온에서 하루 동안 건조시킨다.
마무리	35.	타르트를 틀에서 분리한 후 크림 표면에 크리스탈 스프레이 건 믹스처(73p)를 분사한다.
	36.	물결 띠 모양 초콜릿 장식물을 둘러준다.
	37.	코코넛, 피스타치오, 슈거 페이스트, 원형 초콜릿 디스크, 식용꽃으로 장식해 마무리한다.

SUGAR PASTE	30.	In a mixing bowl, mix powdered sugar, lemon juice, and melted gelatin mass until it forms a ball.
	31.	Roll out the finished paste as thin as possible.
	32.	Cut with desired shape cutters, and dry at an ambient temperature for one day.
LOYAL ICING	33.	Combine powdered sugar, egg whites, and yellow natural food color, and whip.
	34.	Pipe a dot on the dried sugar paste flower decorations.
FINISH	35.	Remove from the mold and spray with crystal spray gun mixture (73p) on the surface of the creams.
	36.	Wrap it with wave shaped band chocolate decoration.
	37.	Decorate with coconut, pistachios, sugar paste, round chocolate disks, and edible flowers to finish.

⁶ PEAR & VANILLA CREAM TARTE

서양배 & 바닐라 크림 타르트

ingredients - ∅ 8cm 12ea

파트 아 사블레(41p) or 스위트 반죽 – 글루텐 프리 (53p)

GLUTEN FREE

클래식 아몬드 크림

버터 149g

설탕 149g

아몬드파우더 149g

전분 19g

달걀전란 113g

바닐라 휩 크림

우유 81g

바닐라빈 1/2개

젤라틴매스 12g

화이트초콜릿 72g

(IVOIRE 35%)

생크림 246g

장식물

다크초콜릿

(EQUATORIALE NOIRE 55%)

아몬드 장식물(155p)

식용금박

기타

서양배 통조림

압솔뤼 크리스탈 글레이즈 100g

물 10g

PATE A SABLÉ(41p) or SWEET DOUGH - GLUTEN FREE(53p)

GLUTEN FREE

CLASSIC ALMOND CREAM

149g Butter

149g Sugar

149g Almond powder

19g Starch

113g Whole eggs

VANILLA WHIPPED CREAM

81g Milk

1/2pc Vanilla bean

12g Gelatin mass

72g White chocolate

(IVOIRE 35%)

246g Fresh cream

DECORATION

Dark chocolate

(EQUATORIALE NOIRE 55%)

Almond decoration(155p)

Edible gold leaves

OTHER

Canned pears

100g Absolu Cristal glaze

10g Water

CLASSIC ALMOND CREAM

1

2

3

4   VANILLA WHIPPED CREAM

5

6

7

8

9-1

9-2

Process

**클래식
아몬드 크림**

1. 버터를 부드럽게 풀어준 후 설탕을 넣고 혼합한다.

2. 아몬드파우더와 전분을 넣고 섞는다.

3. 달걀전란을 조금씩 나눠 넣어가며 혼합한다.

4. 냉장고에서 12시간 휴지시킨 후 사용한다.

바닐라 휩 크림

5. 끓기 직전까지 가열한 우유에 바닐라빈을 넣고 향을 우린 후 젤라틴매스를 혼합한다.

6. 35℃로 녹인 화이트초콜릿에 **5**를 체로 걸러 부어준 후 핸드블렌더로 혼합한다.

7. 35℃로 데운 생크림을 넣고 계속해서 핸드블렌더로 혼합한다.

8. 냉장고에서 12시간 휴지시킨 후 휘핑해 사용한다.

9. 실리콘 몰드(퀸넬)에 채운 후 냉동고에 넣어 단단하게 굳힌다.

**CLASSIC
ALMOND
CREAM**

1. Beat butter to soften and mix with sugar.

2. Mix with almond powder and starch.

3. Gradually mix with whole eggs, a little bit at a time.

4. Rest in the fridge for 12 hours before use.

**VANILLA
WHIPPED
CREAM**

5. Infuse vanilla bean in milk heated until just before boiling. Stir in gelatin mix.

6. Sieve **5** over white chocolate melted to 35℃, then mix using a hand blender.

7. Add fresh cream warmed to 35℃, and continue to mix with a hand blender.

8. Rest in the fridge for 12 hours. Whip to use.

9. Fill into a silicone mold (Quenelle). Freeze completely.

CHOCOLATE DECORATION

11

10

12

13

ASSEMBLY

14

15

16

FINISH

17

18

19

초콜릿 장식물	10.	무스 띠지 위에 템퍼링한 다크초콜릿을 적당량 붓는다.
	11.	스패출러를 이용해 얇게 밀어 편다. 초콜릿이 굳기 시작하면 조심스럽게 무스 띠지를 들어올린다.
	12.	폭 2mm 채칼을 이용해 커팅한다.
	13.	길이 8cm로 커팅한다.

조립	14.	80% 소성한 타르트 셸에 클래식 아몬드 크림을 45g씩 채운다.
	15.	슬라이스한 통조림 서양배를 가지런히 올린다.
	16.	160℃ 오븐에서 약 20분간 굽는다.

마무리	17.	압솔뤼 크리스탈 글레이즈 100g과 물 10g을 냄비에 넣고 끓인 후 서양배 표면에 발라준다.
	18.	타르트 가장자리에 아몬드 장식물을 둘러준다. 몰드에서 빼낸 바닐라 휩 크림 표면에 크리스탈 스프레이 건 믹스처(73p)를 분사한 후 타르트 위에 올려준다.
	19.	초콜릿 장식물과 식용금박을 올려 마무리한다.

CHOCOLATE DECORATION	10.	On a strip of cake collar, pour a moderate amount of tempered dark chocolate.
	11.	Spread thinly using a spatula. When the chocolate starts to crystallize, carefully lift the cake collar.
	12.	Cut using a multi-blade scoring knife with the blades that are 2mm apart.
	13.	Cut into 8cm long strips.

ASSEMBLY	14.	Pipe 45g of classic almond cream into 80% baked tarte shells.
	15.	Slice canned pears and arrange evenly on top.
	16.	Bake for about 20 minutes at 160℃.

FINISH	17.	In a sauce pan, boil 100g of Absolu Cristal glaze and 10g of water, and brush on the surface of the pears.
	18.	Place almond decorations around the edge. Remove frozen vanilla cream from the mold, then spray with crystal spray gun mixture on its surface. Place the cream on top of the tartes.
	19.	Decorate with chocolate decorations and edible gold leaves to finish.

^{7} GRAPEFRUIT TARTE

자몽 타르트

ingredients - ⌀ 12cm 4ea

파트 아 사블레(41p) or
스위트 반죽 – 글루텐 프리
(53p)

GLUTEN
FREE

**자몽과 민트 풍미의
라이트 아몬드 크림**

아몬드파우더 90g

슈거파우더 70g

달걀전란 80g

자몽제스트 자몽 1/3개 분량

애플민트 3g

전분 13g

버터 15g

달걀흰자 24g

설탕 24g

자몽 마멀레이드

자몽 껍질 50g

자몽 과육 200g

설탕 50g

NH펙틴 2.5g

장식물

아몬드 장식물*

허브

아몬드 장식물*

슬라이스 아몬드

달걀흰자

데코스노우

기타

자몽

압솔뤼 크리스탈 글레이즈
100g

물 10g

PATE A SABLÉ(41p) or **SWEET
DOUGH - GLUTEN FREE**(53p)

GLUTEN
FREE

**LIGHT ALMOND CREAM
WITH GRAPEFRUIT AND
MINT AROMA**

90g Almond powder

70g Powdered sugar

80g Whole eggs

Zest of 1/3 grapefruit

3g Applemint

13g Starch

15g Butter

24g Egg whites

24g Sugar

GRAPEFRUIT MARMALADE

50g Grapefruit skin

200g Grapefruit flesh

50g Sugar

2.5g Pectin NH

DECORATION

Almond decoration*

Herbs

ALMOND DECORATION*

Sliced almonds

Egg whites

Decosnow

OTHER

Grapefuits

100g Absolu Cristal glaze

10g Water

LIGHT ALMOND CREAM WITH GRAPEFRUIT
AND MINT AROMA

Process

자몽과 민트 풍미의 라이트 아몬드 크림

1. 볼에 달걀흰자를 넣고 휘핑한다.

2. 설탕을 3~4회 나누어 넣어가며 계속해서 휘핑한다.

3. 거품기로 떠 올렸을 때 머랭 뿔이 탄력이 있으면서도 부드럽게 휘어지는 상태가 될 때까지 휘핑한다.

4. 볼에 달걀전란, 애플민트, 자몽제스트를 넣고 핸드블렌더로 믹싱한다.

5. 아몬드파우더, 슈거파우더를 넣고 휘핑한다.

6. 체 친 전분과 55~60℃로 녹인 버터를 순서대로 혼합한다.

7. 머랭을 2회 나누어 혼합한다.

8. 80% 소성한 타르트 셸에 완성한 자몽과 민트 풍미의 라이트 아몬드 크림을 70g씩 채워준다.

9. 세그먼트한 자몽을 적당량 얹고 160℃ 오븐에서 15분간 굽는다.

LIGHT ALMOND CREAM WITH GRAPEFRUIT AND MINT AROMA

1. Whip egg whites in a mixing bowl.

2. Continue to whip, gradually adding sugar in 3~4 times.

3. Whip until the meringue forms elastic peaks and bends slightly when the whisk is lifted.

4. In a bowl, mix whole eggs, applemint, and grapefruit zest with a hand blender.

5. Whisk in almond powder and powdered sugar, and whip.

6. Mix with sifted starch and then butter melted to 55~60℃ in order.

7. Fold in the meringue in 2 times.

8. Pipe 70g of the light almond cream with grapefruit & mint aroma into 80% baked tarte shells.

9. Place a moderate amount of grapefruit segments on top. Bake for 15 minutes at 160℃.

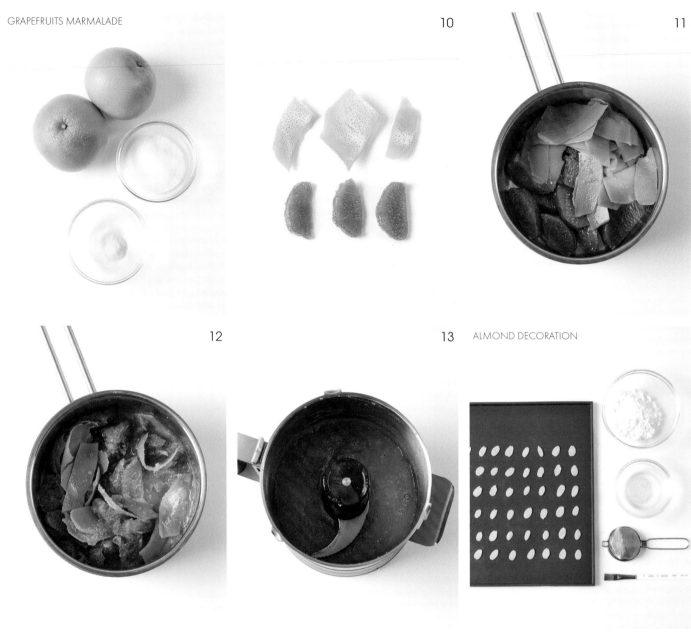

10

11

12

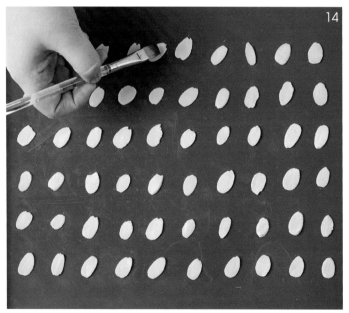

14

15

자몽 마멀레이드

10. 자몽은 껍질과 과육을 분리한다. 자몽 껍질은 끓는 물에 3회 데쳐 쓴맛을 제거한다.

11. 냄비에 자몽 껍질과 자몽 과육을 넣고 가열한다.

12. 45℃가 되면 미리 혼합해둔 설탕과 NH펙틴을 넣고 졸여준다.

13. 푸드프로세서를 이용해 갈아준 후 차갑게 식혀 사용한다.

아몬드 장식물

14. 슬라이스 아몬드 표면에 달걀흰자를 발라준다.

15. 데코스노우를 도포한 후 50℃ 오븐에서 약 2시간 동안 건조시킨다.

GRAPEFRUIT MARMALADE

10. Separate skin and flesh from grapefruits. Blanch the skin three times in boiling water to remove the bitter taste.

11. In a saucepan, heat the grapefruit skin and the grapefruit flesh.

12. When the temperature reaches 45℃, stir in previously mixed sugar and pectin NH and simmer to reduce.

13. Grind in a food processor, and chill to use.

ALMOND DECORATION

14. Brush egg whites on the surface of the sliced almonds.

15. Sift decosnow to cover, then dry in 50℃ oven for about 2 hours.

조립	16.	완전히 식힌 자몽과 민트 풍미의 라이트 아몬드 크림 위에 자몽 마멀레이드를 가득 채운다.
마무리	17.	자몽은 껍질을 벗겨낸 후 세그먼트한다.
	18.	세그먼트한 자몽을 타르트 위에 올려준다.
	19.	압솔뤼 크리스탈 글레이즈 100g과 물 10g을 냄비에 넣고 끓인 후 자몽 표면에 발라준다.
	20.	아몬드 장식물과 허브로 장식해 마무리한다.

ASSEMBLY	16.	Fill completely with grapefruit marmalade over completely cooled light almond cream with grapefruit & mint aroma.
FINISH	17.	Remove the skin from grapefruits, then segment the flesh.
	18.	Place the grapefruit segments on top of the tartes.
	19.	In a saucepan, boil 100g of Absolu Cristal and 10g of water. Brush this mixture on the surface of fruit segments.
	20.	Decorate with almond decorations and herbs to finish.

Nut & Chocolate

FORET NOIRE TARTE

100% CHOCOLATE TARTE

PISTACHIO & CHERRY TARTE

BLACK SESAME & HAZELNUT TARTE

⁸ FORET NOIRE TARTE

포레누아 타르트

ingredients - ⌀ 16cm 3ea

초콜릿 스트로이젤

버터 94g
설탕 94g
소금 0.6g
아몬드파우더 118g
박력분 82g
카카오파우더 12g

초콜릿 비스퀴

달걀노른자 76g
박력분 15g
카카오파우더 31g
버터 7g
다크초콜릿 38g
(🍫 CARAIBE 66%)
달걀흰자 107g
전화당 107g

사워체리 젤리

체리 퓌레 357g
설탕 28g
전분 15g
젤라틴매스 20.1g

장식물

다크초콜릿
(🍫 EQUATORIALE NOIRE 55%)
체리
허브

바닐라 휩 크림(147p)

우유 89g
바닐라빈 2/3개
젤라틴매스 13.1g
화이트초콜릿 79g
(🍫 IVOIRE 35%)
샹크림 269g

다크초콜릿 휩 크림

우유 82g
젤라틴매스 13.1g
다크초콜릿 88g
(🍫 MANJARI 64%)
샹크림 269g

CHOCOLATE STREUSEL

94g Butter
94g Sugar
0.6g Salt
118g Almond powder
82g Cake flour
12g Cacao powder

CHOCOLATE BISCUIT

76g Egg yolks
15g Cake flour
31g Cacao powder
7g Butter
38g Dark chocolate
(🍫 CARAIBE 66%)
107g Egg whites
107g Inverted sugar

SOUR CHERRY JELLY

357g Cherry purée
28g Sugar
15g Starch
20.1g Gelatin mass

DECORATION

Dark chocolate
(🍫 EQUATORIALE NOIRE 55%)
Cherries
Herbs

VANILLA WHIPPED CREAM(147p)

89g Milk
2/3pc Vanilla bean
13.1g Gelatin mass
79g White chocolate
(🍫 IVOIRE 35%)
269g Fresh cream

DARK CHOCOLATE WHIPPED CREAM

82g Milk
13.1g Gelatin mass
88g Dark chocolate
(🍫 MANJARI 64%)
269g Fresh cream

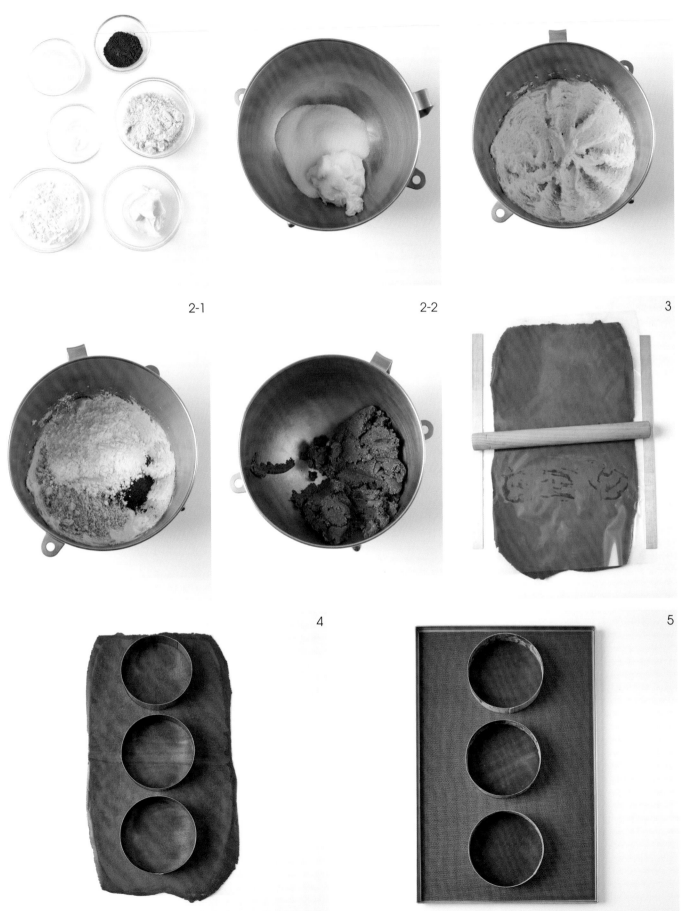

Process

초콜릿
스트로이젤

1. 버터를 부드럽게 풀어준 후 설탕과 소금을 넣고 혼합한다.

2. 아몬드파우더, 카카오파우더, 박력분을 넣고 한 덩어리가 될 때까지 믹싱한다.

3. 두 장의 투명 필름 사이에 반죽을 놓고 2mm 두께로 밀어 편다. 이때 2mm 두께의 바를 이용하면 일정한 두께로 밀어 펼 수 있다.

4. 지름 16cm 원형 틀을 이용해 커팅한다.

5. 틀 째 오븐에 넣고 160℃에서 12분간 굽는다.

CHOCOLATE
STREUSEL

1. Beat butter to soften, then mix with sugar and salt.

2. Add almond powder, cacao powder, and cake flour. Mix until it forms a ball.

3. Place the dough between two sheets of clear plastic film and roll to 2mm thickness. Using 2mm thick bars will help to roll out evenly.

4. Cut using 16cm diameter ring molds.

5. Place the streusels with the molds in a 160℃ oven, and bake for 12 minutes.

초콜릿 비스퀴

6. 박력분과 카카오파우더를 함께 체 친다.

7. 다크초콜릿과 55~60℃로 녹인 버터를 혼합한다.

8. 볼에 달걀흰자를 넣고 휘핑한다.

9. 전화당을 넣고 계속해서 휘핑한다.

10. 거품기로 떠 올렸을 때 머랭 뿔이 탄력이 있으면서도 부드럽게 휘어지는 상태가 될 때까지 휘핑한다.

11. 머랭에 달걀노른자를 넣고 섞어준다.

12. 체 친 박력분과 카카오파우더를 혼합한다.

13. **7**에 반죽 일부를 넣고 혼합한다.

14. 다시 전체 반죽에 넣어 혼합한다.

15. 초콜릿 스트로이젤 위에 완성된 초콜릿 비스퀴 반죽을 120g씩 팬닝한다.

16. 스패출러로 표면을 고르게 정리한 후 160℃ 오븐에서 15분간 굽는다.

CHOCOLATE BISCUIT

6. Sift cake flour and cacao powder together.

7. Mix with dark chocolate and butter melted to 55~60℃.

8. In a mixing bowl, whip egg whites.

9. Add inverted sugar, and continue whipping.

10. Whip until the meringue forms elastic peaks and bends slightly when the whisk is lifted.

11. Stir egg yolks in the meringue.

12. Mix with sifted cake flour and cacao powder.

13. Mix part of the batter into the butter mixture(**7**).

14. Combine the rest.

15. On top of the finished chocolate streusel, pipe 120g of chocolate biscuit batter.

16. Even out the surface with a spatula. Bake for 15 minutes at 160℃.

SOUR CHERRY JELLY

17

18

19

20

21

DARK CHOCOLATE WHIPPED CREAM

22

23

24

사워체리 젤리

17. 냄비에 체리 퓌레를 넣고 45℃로 가열한다.

18. 미리 섞어둔 설탕과 전분을 넣고 전분이 충분히 호화될 때까지 가열한다.

19. 젤라틴매스를 넣고 섞어준 후 핸드블렌더를 이용해 균일하게 혼합한다.

20. 초콜릿 비스퀴 위에 완성한 사워체리 젤리를 130g씩 부어준다.

21. 냉동고에 넣어 단단하게 굳혀준다.

**다크초콜릿
휩 크림**

22. 끓기 직전까지 가열한 우유에 젤라틴매스를 혼합한다. 35℃로 녹인 다크초콜릿에 붓고 핸드블렌더로 혼합한다.

23. 35℃로 데운 생크림을 넣고 계속해서 핸드블렌더를 이용해 혼합한다.

24. 냉장고에서 12시간 휴지시킨 후 휘핑해 사용한다.

**SOUR CHERRY
JELLY**

17. In a saucepan, heat cherry purée to 45℃.

18. Stir in previously mixed sugar and starch mixture, and continue to heat until gelatinized enough.

19. Stir in gelatin mass and mix evenly with a hand blender.

20. On top of the chocolate biscuit, pour 130g of the finished sour cherry jelly.

21. Freeze completely.

**DARK
CHOCOLATE
WHIPPED CREAM**

22. Stir gelatin mass into mik heated until just before boiling. Pour over dark chocolate melted to 35℃, and mix with a hand blender.

23. Add fresh cream warmed to 35℃, and continue to mix with a hand blender.

24. Rest in the fridge for 12 hours. Whip to use.

CHOCOLATE DECORATION ❶

25

26

27

28

29

CHOCOLATE DECORATION ❷   30

ASSEMBLY

31

FINISH   32

33

34

초콜릿 장식물 ❶	25.	무스 띠지를 55cm 길이로 잘라 준비한 후 템퍼링한 다크초콜릿을 적당량 붓는다.
	26.	스패출러를 이용해 얇게 밀어 편다.
	27.	초콜릿이 굳기 시작하면 조심스럽게 무스 띠지를 들어올린다.
	28.	3cm 폭으로 커팅한다.
	29.	16cm 무스 틀에 감아준다.
초콜릿 장식물 ❷	30.	139p 초콜릿 장식물 ❷ 방식을 참고하여 삼각형 모양 장식물을 만든다.
조립	31.	사워체리 젤리 위에 바닐라 휩 크림과 다크초콜릿 휩 크림을 번갈아가며 파이핑한다.
		* 원형(지름 1cm) 깍지를 이용해 원뿔 모양으로 파이핑한다.
		냉동고에서 단단하게 굳힌다.
마무리	32.	타르트를 틀에서 분리한 후 크림 표면에 크리스탈 스프레이 건 믹스처(73p)를 분사한다.
	33.	띠 모양의 초콜릿 장식물을 둘러준다.
	34.	삼각형으로 재단한 초콜릿 장식물, 체리, 허브를 올려 마무리한다.

CHOCOLATE DECORATION ❶	25.	Prepare cake collar cut to 55cm. Pour a moderate amount of tempered dark chocolate.
	26.	Spread thinly using a spatula.
	27.	When the chocolate starts to crystallize, lift the cake collar carefully.
	28.	Cut into 3cm wide strips.
	29.	Wrap around 16cm ring molds.
CHOCOLATE DECORATION ❷	30.	Refer to chocolate decoration ❷ on page 139 to make triangle-shaped decorations.
ASSEMBLY	31.	On top of the sour cherry jelly, pipe vanilla whipped cream and dark chocolate whipped cream alternately.
		* Use a round tip (1cm diameter) to pipe in cone shapes.
		Freeze completely.
FINISH	32.	Remove the tartes from the molds. Spray crystal spray gun mixture (73p) on the surface of the creams.
	33.	Wrap with the chocolate strip decoration.
	34.	Decorate with chocolate decorations cut into triangles, cherries, and herbs to finish.

9 100% CHOCOLATE TARTE

100% 초콜릿 타르트

ingredients - Ø 7cm 12ea

파트 아 사블레 – 초콜릿(45p)

헤이즐넛 카카오 프랄리네

헤이즐넛 250g
카카오닙 25g
설탕 250g
물 58g
게랑드소금 2.5g

초콜릿 비스퀴

달걀노른자 128g
박력분 26g
카카오파우더 51g
버터 13g
다크초콜릿 64g
(CARAIBE 66%)
달걀흰자 180g
전화당 180g

기타

게랑드소금
데코스노우

비터 초콜릿 크림

달걀노른자 75g
설탕 38g
우유 103g
생크림 103g
다크초콜릿 80g
(CARAIBE 66%)

카카오 밀크*

카카오닙 16g
우유 80g

카카오 휩 크림

카카오 밀크* 73g
젤라틴매스 12g
다크초콜릿 80g
(MANJARI 64%)
생크림 246g

장식물

다크초콜릿
(EQUATORIALE NOIRE 55%)
카카오파우더
식용금박

글레이즈

물 75g
설탕 150g
물엿 150g
연유 100g
젤라틴매스 60g
다크초콜릿 150g
(CARAIBE 66%)
붉은색 천연 식용 색소
적당량

PATE A SABLÉ - CHOCOLATE(45p)

HAZELNUT CACAO PRALINE

250g Hazelnuts
25g Cacao nips
250g Sugar
58g Water
2.5g Guérande salt

CHOCOLATE BISCUIT

128g Egg yolks
26g Cake flour
51g Cacao powder
13g Butter
64g Dark chocolate
(CARAIBE 66%)
180g Egg whites
180g Inverted sugar

OTHER

Guérande salt
Decosnow

BITTER CHOCOLATE CREAM

75g Egg yolks
38g Sugar
103g Milk
103g Fresh cream
80g Dark chocolate
(CARAIBE 66%)

CACAO MILK *

16g Cacao nips
80g Milk

CACAO WHIPPED CREAM

73g Cacao milk *
12g Gelatin mass
80g Dark chocolate
(MANJARI 64%)
246g Fresh cream

DECORATION

Dark Chocolate
(EQUATORIALE NOIRE 55%)
Cacao powder
Edible gold leaves

GLAZE

75g Water
150g Sugar
150g Corn syrup
100g Condensed milk
60g Gelatin mass
150g Dark chocolate
(CARAIBE 66%)
QS Red natural food
color

Process

헤이즐넛 카카오 프랄리네

1. 헤이즐넛은 140℃로 예열된 오븐에서 약 8분간 구워준다. 냄비에 물과 설탕을 넣고 118~121℃까지 끓인다.

2. 118~121℃가 되면 불에서 내린 후 따뜻한 상태의 헤이즐넛을 넣고 섞어준다.

3. 시럽이 하얗게 재결정화 상태가 될 때까지 계속해서 저어준다.

4. 다시 불에 올려 골고루 저어주며 계속해서 가열한다.

5. 캐러멜화가 시작되면 카카오닙을 넣고 진한 갈색이 날 때까지 조금 더 진행한다.

6. 실팻에 넓게 펼쳐 완전히 식힌다.

7. 푸드프로세서에 식힌 캐러멜라이즈 헤이즐넛과 게랑드소금을 넣고 갈아준다.

8. 흐르는 정도의 페이스트 상태가 되면 마무리한다.

HAZELNUT CACAO PRALINE

1. Roast hazelnuts for about 8 minutes in an oven preheated to 140℃. In a saucepan, boil water and sugar until 118~121℃.

2. When the temperature reaches 118~121℃, remove from heat and stir in warmed hazelnuts.

3. Continue to stir until the sugar recrystallizes and turns white.

4. Place it back over the heat, and continue to stir.

5. When it starts to caramelize, add cacao nips and continue to cook until the color darkens a bit more.

6. Spread over a Silpat, and cool completely.

7. Grind cooled caramelized hazelnuts and Guérande salt in a food processor.

8. Finish when the texture turns into a fluid paste.

초콜릿 비스퀴

9. 박력분과 카카오파우더를 함께 체 친다.

10. 다크초콜릿과 55~60℃로 녹인 버터를 혼합한다.

11. 볼에 달걀흰자를 넣고 휘핑한다.

12. 전화당을 넣고 계속해서 휘핑한다.

13. 거품기로 떠 올렸을 때 머랭 뿔이 탄력이 있으면서도 부드럽게 휘어지는 상태가 될 때까지 휘핑한다.

14. 머랭에 달걀노른자를 넣고 섞어준다.

15. 체 친 박력분과 카카오파우더를 혼합한다.

16. **10**에 반죽 일부를 넣고 혼합한다.

17. 다시 전체 반죽에 넣어 혼합한다.

18. 32.5×32.5cm, 높이 1cm 실리콘 몰드에 완성된 초콜릿 비스퀴 반죽을 팬닝한 후 스패출러로 표면을 고르게 정리해 160℃ 오븐에서 15분간 굽는다.

19. 비스퀴가 완전히 식으면 껍질이 달라붙지 않도록 비스퀴 표면에 데코스노우를 소량 뿌려준 후 뒤집어 몰드를 제거한다.

20. 지름 5.5cm 원형 커터로 커팅한다.

CHOCOLATE BISCUIT

9. Sift cake flour and cacao powder together.

10. Mix with dark chocolate and butter melted to 55~60℃.

11. In a mixing bowl, whip egg whites.

12. Add inverted sugar and continue whipping.

13. Whip until the meringue forms elastic peaks and bends slightly when the whisk is lifted.

14. Stir egg yolks in the meringue.

15. Mix with sifted cake flour and cacao powder.

16. Mix part of the batter into the butter mixture(**10**).

17. Combine with the remaining batter.

18. In a silicone mold, 32.5 × 32.5 × 1cm (height), pour chocolate biscuit batter and spread evenly with a spatula. Bake for 15 minutes at 160℃.

19. After the biscuit is completely cooled, sprinkle a little bit of decosnow on the surface to prevent the skin from sticking, then turn it upside down to remove the mold.

20. Cut using a 5.5cm diameter round cutter.

BITTER CHOCOLATE CREAM

21

22

23

24

25

CACAO MILK

26

27

28

비터 초콜릿 크림

21. 볼에 달걀노른자, 설탕 1/2을 넣고 혼합한다. 냄비에 생크림, 우유, 나머지 설탕을 넣고 45℃로 데운 후 달걀노른자 믹스처에 조금씩 부어가며 혼합한다.

22. 냄비에 옮겨 83~85℃로 가열해 크렘 앙글레이즈를 만든다.

23. 체에 걸러준 후 45℃까지 식힌다.

24. 35℃로 녹인 다크초콜릿에 **23**을 3~4회 나누어 섞는다.

25. 핸드블렌더로 혼합한 후 냉장고에서 12시간 휴지시켜 사용한다.

카카오 밀크

26. 160℃ 오븐에서 3분간 구운 카카오닙과 우유를 수비드 백에 담는다.

27. 약 60℃로 1시간 동안 수비드해 카카오 향을 추출한다.

28. 체에 걸러 카카오 밀크를 추출한다.

BITTER CHOCOLATE CREAM

21. In a bowl, mix egg yolks and half of the sugar. In a saucepan, heat fresh cream, milk, and remaining sugar to 45℃. Gradually mix into the egg yolk mixture, a little bit at a time.

22. Pour into a saucepan and boil until 83~85℃ to make creme anglaise.

23. Strain and cool it down until 45℃.

24. Gradually mix **23** to dark chocolate melted to 35℃, in 3~4 times.

25. Mix with a hand blender. Rest in the fridge for 12 hours before use.

CACAO MILK

26. In a sous vide bag, pour milk and cacao nips baked for 3 minutes at 160℃.

27. Sous vide at 60℃ for 1 hour to extract cacao aroma.

28. Strain to extract cacao milk.

CACAO WHIPPED CREAM

GLAZE

카카오
휩 크림

29. 60℃로 데운 카카오 밀크에 젤라틴매스를 넣고 혼합한다.

30. 35℃로 녹인 다크초콜릿에 **29**를 조금씩 나누어 섞어준 후 핸드블렌더로 혼합한다.

31. 35℃로 데운 생크림을 넣고 계속해서 핸드블렌더로 혼합한다.

32. 냉장고에 12시간 휴지시킨 후 휘핑해 사용한다.

글레이즈

33. 냄비에 물과 설탕을 넣고 가열한다.

34. 시럽 상태가 되면 물엿을 넣고 103℃까지 가열한 후 불에서 내린다.

35. 연유와 젤라틴매스를 순서대로 넣고 섞는다.

36. 35℃로 녹인 다크초콜릿과 붉은색 천연 식용 색소에 **35**를 붓고 핸드블렌더로 혼합한다.

37. 사용할 때는 다시 35℃로 온도를 맞춘 다음 핸드블렌더로 균일하게 믹싱해 사용한다.

CACAO
WHIPPED
CREAM

29. Stir gelatin mass in the cacao milk heated to 60℃.

30. Stir **29** into dark chocolate melted to 35℃, a little bit at a time. Combine with a hand blender.

31. Add fresh cream warmed to 35℃, and continue to mix with a hand blender.

32. Rest in the fridge for 12 hours. Whip to use.

GLAZE

33. In a saucepan, heat water and sugar.

34. When it turns into syrup, add corn syrup and continue to boil until 103℃. Remove from heat.

35. Stir in condensed milk and gelatin mass in order.

36. Pour **35** over dark chocolate melted to 35℃ and red natural food color, and mix using a hand blender.

37. To use, reset the glaze to 35℃ and mix thoroughly using a hand blender.

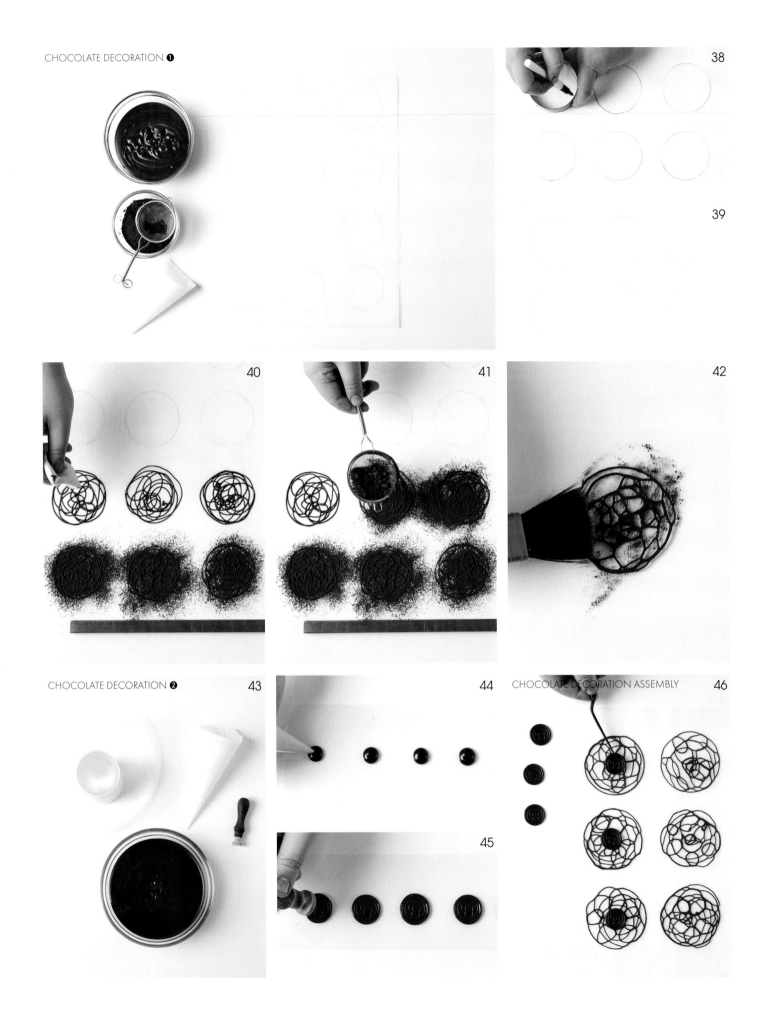

38

39

40

41

42

43

44

45

CHOCOLATE DECORATION ASSEMBLY 46

초콜릿 장식물 ❶	38.	종이에 지름 7cm 원형 틀을 이용해 라인을 그린다.
	39.	가이드 종이 위에 베이킹 페이퍼를 올린다.
	40.	템퍼링한 다크초콜릿을 불규칙한 모양으로 파이핑한다.
	41.	초콜릿이 완전히 굳기 전에 카카오파우더를 도포한다.
	42.	초콜릿이 완전히 굳으면 붓으로 여분의 카카오파우더를 털어낸다.
초콜릿 장식물 ❷	43.	사용할 도장은 미리 냉동고에 넣어두어 차가운 상태로 준비한다.
	44.	무스 띠지에 템퍼링한 다크초콜릿을 지름 1cm 사이즈로 파이핑한다.
	45.	차가운 상태의 도장으로 파이핑한 초콜릿을 눌러 문양을 만든다.
초콜릿 장식물 조립	46.	완성한 초콜릿 장식물 ❶의 중앙에 템퍼링한 다크초콜릿을 소량 파이핑한 후 초콜릿 장식물 ❷를 고정시킨다.

CHOCOLATE DECORATION ❶	38.	On a paper, draw 7cm diameter rounds using a ring mold as a guide.
	39.	Place parchment paper on top of the guide paper(38).
	40.	Pipe tempered dark chocolate in irregular shape within the line.
	41.	Sift cacao powder before the chocolate sets.
	42.	After the chocolates are completely crystallized, brush off the excess cacao powder.
CHOCOLATE DECORATION ❷	43.	Store the chocolate stamp of your choice in the freezer in advance to keep it cold.
	44.	On a cake collar, pipe tempered dark chocolate in 1cm diameter.
	45.	Press on the piped chocolate drops with the cold stamp to make imprints.
CHOCOLATE DECORATION ASSEMBLY	46.	Pipe a small amount of tempered dark chocolate on the center of finished chocolate decoration ❶, and fix chocolate decoration ❷.

조립	47.	100% 소성한 타르트 셸에 헤이즐넛 카카오 프랄리네를 18g씩 파이핑한다.
	48.	게랑드소금을 뿌려준다.
	49.	초콜릿 비스퀴를 올려준 후 평평한 도구로 가볍게 눌러 평탄 작업을 한다.
	50.	비터 초콜릿 크림을 30g씩 채워준다.
	51.	냉장고에서 3시간 휴지시킨 후 냉동고에 넣어 단단하게 굳힌다.
마무리	52.	타르트 표면에 글레이즈를 채워준다.
	53.	원형(지름 1cm) 깍지를 이용해 카카오 휩 크림을 원뿔 모양으로 파이핑한다.
	54.	초콜릿 장식물과 식용금박을 올려 마무리한다.

ASSEMBLY	47.	Pipe18g of hazelnut cacao praline in the 100% baked tarte shells.
	48.	Sprinkle Guérande salt.
	49.	Place chocolate biscuits and lightly press the top with a flat tool.
	50.	Fill with 30g of bitter chocolate cream.
	51.	Rest in the fridge for 3 hours, then freeze completely.
FINISH	52.	Fill the surface of the tartes with the glaze.
	53.	Pipe cacao whipped cream in cone shape using a round tip (1cm diameter).
	54.	Decorate with chocolate decorations and edible gold leaves to finish.

¹⁰ PISTACHIO & CHERRY TARTE

피스타치오 & 체리 타르트

ingredients - ⌀ 7cm 12ea

파트 아 사블레(41p) or 스위트 반죽 – 글루텐 프리(53p)

GLUTEN FREE

피스타치오 페이스트*

피스타치오 252g
설탕 126g
물 42g
게랑드소금 2.5g

피스타치오 풍미의 라이트 아몬드 크림

아몬드파우더 90g
슈거파우더 70g
달걀전란 80g
피스타치오 페이스트* 25g
전분 13g
버터 15g
달걀흰자 24g
설탕 24g

사워체리 젤*

물 35g
사워체리 퓌레 84g
레몬주스 50g
아가아가 2.5g
설탕 28g

장식물

화이트초콜릿
(OPALYS 33%)
체리
데코젤
식용금박

피스타치오 바바로아

달걀노른자 54g
설탕 21g
피스타치오 페이스트* 27g
우유 95g
바닐라빈 1/2개
젤라틴매스 25.2g
생크림 176g

초콜릿 스프레이 건 믹스처

화이트초콜릿 90g
(OPALYS 33%)
카카오버터 90g
노란색 천연 식용 색소
녹색 천연 식용 색소

사워체리 인서트

사워체리 젤* 143g
체리 207g

PATE A SABLÉ(41p) or SWEET DOUGH - GLUTEN FREE(53p)

GLUTEN FREE

PISTACHIO PASTE*

252g Pistachios
126g Sugar
42g Water
2.5g Guérande salt

LIGHT ALMOND CREAM WITH PISTACHIO AROMA

90g Almond powder
70g Powdered sugar
80g Whole eggs
25g Pistachio paste*
13g Starch
15g Butter
24g Egg whites
24g Sugar

SOUR CHERRY GEL*

35g Water
84g Sour cherry purée
50g Lemon juice
2.5g Agar-agar
28g Sugar

DECORATION

White chocolate
(OPALYS 33%)
Cherries
Decogel
Edible gold leaves

PISTACHIO BAVAROIS

54g Egg yolks
21g Sugar
27g Pistachio paste*
95g Milk
1/2pc Vanilla bean
25.2g Gelatin mass
176g Fresh cream

CHOCOLATE SPRAY GUN MIXTURE

90g White chocolate
(OPALYS 33%)
90g Cacao butter
Yellow natural food color
Green natural food color

SOUR CHERRY INSERT

143g Sour cherry gel*
207g Cherries

Process

**피스타치오
페이스트**

1. 피스타치오는 170℃로 예열된 오븐에서 약 15분간 구워준다. 냄비에 물과 설탕을 넣고 118~121℃까지 끓인다.

2. 118~121℃가 되면 불에서 내린 다음 따뜻한 상태의 피스타치오를 넣고 섞어준다.

3. 시럽이 하얗게 재결정화 상태가 될 때까지 계속해서 저어준다.

4. 실팻에 넓게 펼친 후 완전히 식힌다.

5. 푸드프로세서에 식힌 피스타치오와 게랑드소금을 넣고 갈아준다.

6. 흐르는 정도의 페이스트 상태가 되면 마무리한다.

**PISTACHIO
PASTE**

1. Preheat oven to 170°C. Roast pistachios for about 15 minutes. In a saucepan, boil water and sugar until the temperature reaches 118~121°C.

2. Once it reaches 118~121°C, remove from heat and mix with the warm pistachios.

3. Continue to stir until the syrup recrystallizes and turns white.

4. Spread over Silpat to cool completely.

5. In a food processor, grind cooled pistachios and Guérande salt.

6. Finish when the mixture turns into a fluid paste.

LIGHT ALMOND CREAM WITH
PISTACHIO AROMA

**피스타치오 풍미의
라이트
아몬드 크림**

7. 볼에 달걀흰자를 넣고 휘핑한다.

8. 설탕을 3~4회 나누어 넣어가며 계속해서 휘핑한다.

9. 거품기로 떠 올렸을 때 머랭 뿔이 탄력이 있으면서도 부드럽게 휘어지는 상태가 될 때까지 휘핑한다.

10. 달걀전란, 아몬드파우더, 슈거파우더를 혼합한 후 휘핑한다.

11. 피스타치오 페이스트를 넣고 혼합한다.

12. 체 친 전분과 55~60℃로 녹인 버터를 순서대로 넣고 섞는다.

13. 머랭을 2회 나누어 혼합한다.

14. 80% 소성한 타르트 셸에 완성한 피스타치오 풍미의 라이트 아몬드 크림을 20g씩 채워준 후
    160℃ 오븐에서 15분간 굽는다.

**LIGHT ALMOND
CREAM WITH
PISTACHIO
AROMA**

7. Whip egg whites in a mixing bowl.

8. Gradually add sugar in 3~4 times, and continue whipping.

9. Whip until the meringue forms elastic peaks and bends slightly when the whisk is
    lifted.

10. Stir in whole eggs, almond powder, and powdered sugar, and whip.

11. Mix with pistachio paste.

12. Combine sifted starch and butter melted to 55~60℃ in order.

13. Fold in the meringue in 2 times.

14. Pipe 20g of finished light almond cream with pistachio aroma in each 80% baked
    tarte shells and bake for 15 minutes at 160℃.

SOUR CHERRY GEL

15

16

17 SOUR CHERRY INSERT

18

WHITE CHOCOLATE DISC

19

20

21

사워체리 젤

15. 냄비에 물, 사워체리 퓌레, 레몬주스를 넣고 45℃로 데운다.

16. 미리 섞어둔 설탕과 아가아가를 넣고 끓어오를 때까지 가열한 후 볼에 옮겨 냉장고에서 굳힌다.

17. 핸드블렌더를 이용해 부드러운 젤 상태가 될 때까지 갈아준다.

사워체리 인서트

18. 사워체리 젤과 사방 3mm로 작게 자른 체리를 섞어 사워체리 인서트를 완성한다.

**화이트초콜릿
디스크**

19. 두 장의 투명 필름 사이에 템퍼링한 화이트초콜릿을 적당량 부어준다.

20. 필름을 덮고 밀대를 이용해 균일한 두께로 밀어 편다.

21. 지름 6.3cm 원형 커터로 커팅한다.

**SOUR
CHERRY GEL**

15. In a saucepan, heat water, sour cherry purée, and lemon juice to 45℃.

16. Whisk in previously mixed sugar and agar-agar, and heat until the mixture boils. Pour into a bowl and store in the fridge to set.

17. Once it's set, blend with a hand blender to make a soft gel.

**SOUR CHERRY
INSERT**

18. Mix sour cherry gel and cherries cut to 3mm cubes to make sour cherry insert.

**WHITE
CHOCOLATE
DISC**

19. Between two sheets of clear plastic film pour a moderate amount of tempered white chocolate.

20. Cover with the film and roll to even thickness using a rolling pin.

21. Cut with a 6.3cm diameter round cutter.

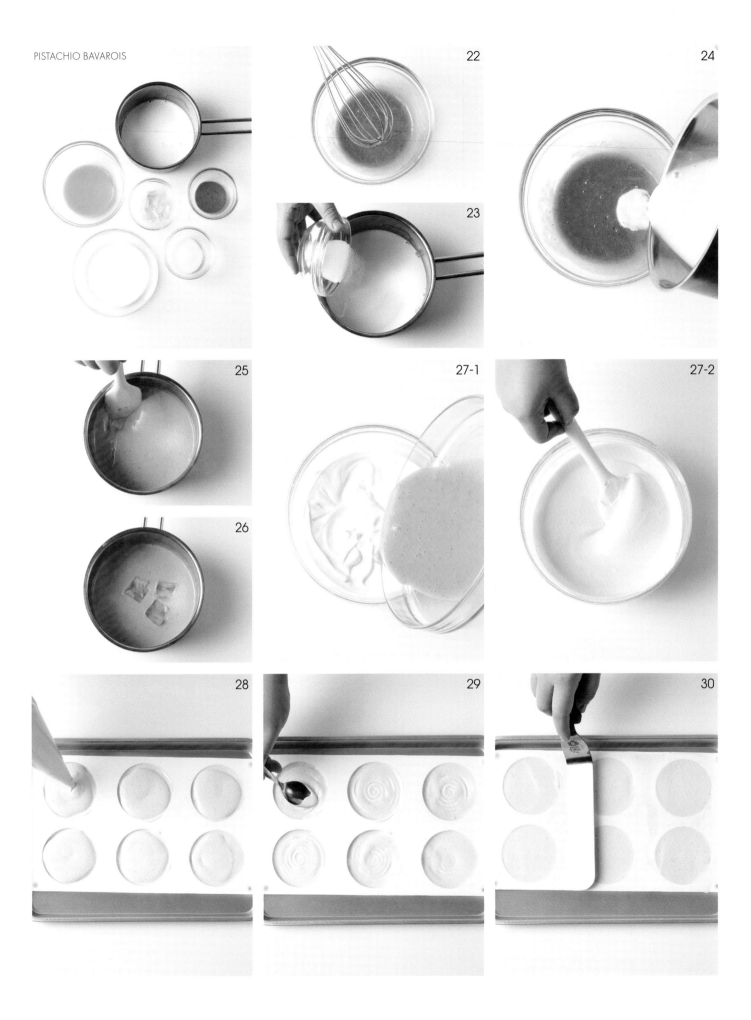

**피스타치오
바바로아**

22. 볼에 달걀노른자, 설탕 1/2, 피스타치오 페이스트를 넣고 혼합한다.

23. 냄비에 우유, 나머지 설탕, 바닐라빈을 넣고 끓기 직전까지 가열한 후 바닐라 향을 우린다.

24. **22**에 **23**을 조금씩 부어가며 혼합한다.

25. 다시 냄비로 옮겨 83~85℃로 가열해 피스타치오 크렘 앙글레이즈를 만든다.

26. 젤라틴매스를 넣고 혼합한 후 25℃까지 식힌다.

27. 무스 상태로 휘핑한 생크림과 혼합한다.

28. 지름 7.3cm 투르빌리옹 실리콘 몰드에 완성한 피스타치오 바바로아를 채워준다.

29. 스푼을 이용해 몰드의 굴곡진 부분까지 골고루 채워지도록 작업한다.

30. 스패출러로 표면을 고르게 정리한 후 냉동고에서 단단하게 굳힌다.

**PISTACHIO
BAVAROIS**

22. In a bowl, mix egg yolks, 1/2 of sugar, and pistachio paste.

23. In a saucepan, heat milk, remaining sugar, and vanilla bean until just before boiling to infuse the vanilla aroma.

24. Gradually stir **23** into **22**.

25. Pour back into a saucepan and heat until 83~85℃ to make pistachio creme anglaise.

26. Stir in gelatin mass and cool down to 25℃.

27. Mix with fresh cream whipped to a mousse consistency.

28. Pour finished pistachio bavarois in 7.3cm diameter Tourbillon silicone molds.

29. Make sure to fill all the ridges of the mold evenly with a spoon.

30. Even out the surface with a spatula, then freeze completely.

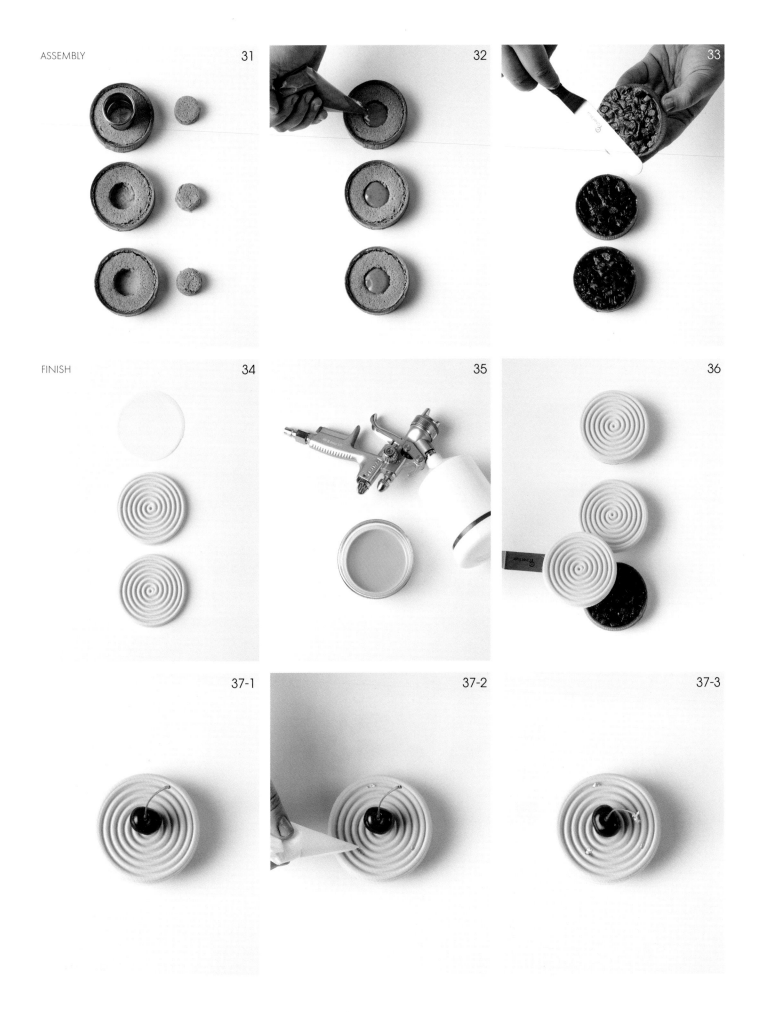

조립	31.	지름 2.5cm 원형 커터를 이용해 완전히 식힌 피스타치오 풍미의 라이트 아몬드 크림 중심부에 구멍을 뚫어 공간을 만든다. 이때 타르트 셸까지 뚫지 않도록 주의한다.
	32.	만든 공간 안에 피스타치오 페이스트를 가득 채운 후 냉동고에서 단단하게 굳힌다.
	33.	사워 체리 인서트를 타르트 셸 높이까지 가득 채운다.
마무리	34.	화이트초콜릿 디스크에 피스타치오 바바로아를 얹어준다.
	35.	초콜릿 스프레이 건 믹스처를 분사한다.
	36.	타르트 셸 위에 올려준다.
	37.	체리, 데코젤, 식용금박으로 장식해 마무리한다.

ASSEMBLY	31.	In the center of completely cooled light almond cream with pistachio aroma, make a hole using a 2.5cm diameter round cutter. Be careful not to cut through the tarte shell on the very bottom.
	32.	Fill up with pistachio paste in the hole, then freeze completely.
	33.	Pour sour cherry insert until the edge of the tarte shell.
FINISH	34.	Place pistachio bavarois on the white chocolate disc.
	35.	Spray with chocolate spray gun mixture.
	36.	Place on top of the tarte.
	37.	Decorate with cherries, decogel, and edible gold leaves to finish.

11 BLACK SESAME & HAZELNUT TARTE

흑임자 & 헤이즐넛 타르트

ingredients - Ø 8cm 12ea

세사미 스트로이젤

버터 96g
카소나드 96g
소금 0.6g
헤이즐넛파우더 119g
박력분 95g
참깨 47g

**흑임자 풍미의
라이트 헤이즐넛 크림**

헤이즐넛파우더 132g
슈거파우더 103g
달걀전란 117g
오렌지제스트 오렌지
2/3개 분량
흑임자 페이스트 37g
전분 19g
버터 22g
달걀흰자 35g
설탕 35g

**헤이즐넛 세사미
프랄리네***

헤이즐넛 250g
흑임자 20g
참깨 20g
설탕 250g
물 58g
게랑드소금 2.5g

**헤이즐넛 세사미
프랄리네 크림**

달걀노른자 50g
설탕 33g
우유 93g
생크림 93g
블론드초콜릿 82g
(⚜ DULCEY 32%)
카카오버터 15g
헤이즐넛 세사미
프랄리네* 32g
젤라틴매스 9g

블론드 휩 크림

젤라틴매스 12g
우유 84g
블론드초콜릿 96g
(⚜ DULCEY 32%)
생크림 219g

장식물

밀크초콜릿
(⚜ JIVARA LATTE 40%)
헤이즐넛
식용금박
캐러멜 소스(265p)

SESAME STREUSEL

96g Butter
96g Cassonade
0.6g Salt
119g Hazelnut
powder
95g Cake flour
47g Sesame seeds

**LIGHT HAZELNUT
CREAM WITH BLACK
SESAME AROMA**

132g Hazelnut powder
103g Powdered sugar
117g Whole eggs
Zest of 2/3 orange
37g Black sesame
paste
19g Starch
22g Butter
35g Egg whites
35g Sugar

**HAZELNUT SESAME
PRALINE***

250g Hazelnuts
20g Black sesame
seeds
20g Sesame seeds
250g Sugar
58g Water
2.5g Guérande salt

**HAZELNUT SESAME
PRALINE CREAM**

50g Egg yolks
33g Sugar
93g Milk
93g Fresh cream
82g Blond Chocolate
(⚜ DULCEY 32%)
15g Cacao butter
32g Hazelnut Sesame
Praline*
9g Gelatin mass

**BLOND WHIPPED
CREAM**

12g Gelatin mass
84g Milk
96g Blond chocolate
(⚜ DULCEY 32%)
219g Fresh cream

DECORATION

Milk chocolate
(⚜ JIVARA LATTE 40%)
Hazelnuts
Edible gold leaves
Caramel sauce(265p)

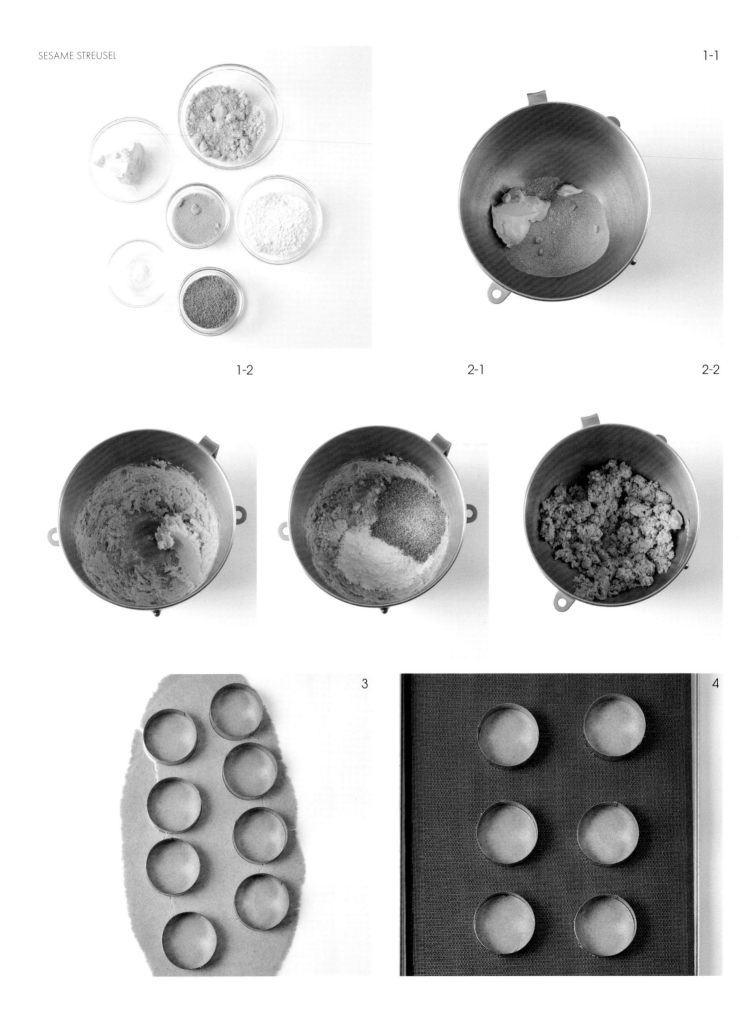

1-1

1-2

2-1

2-2

3

4

Process

**세사미
스트로이젤**

1. 버터를 부드럽게 풀어준 후 카소나드와 소금을 넣고 혼합한다.

2. 헤이즐넛파우더, 참깨, 박력분을 넣고 한 덩어리가 될 때까지 믹싱한다.

3. 2mm 두께로 밀어 편 후 냉동고에서 단단히 굳혀준다.

4. 지름 8cm 원형 틀로 커팅한 후 틀 째 오븐에 넣고 160℃에서 10분간 굽는다.

**SESAME
STREUSEL**

1. Beat butter to soften, and mix with cassonade and salt.

2. Mix with hazelnut powder, sesame seeds, and cake flour until it forms a ball.

3. Roll out to 2mm thickness, and store in the freezer to harden.

4. Cut with 8cm diameter ring molds. Bake with the molds for 10 minutes at 160℃.

LIGHT HAZELNUT CREAM WITH BLACK
SESAME AROMA

**흑임자 풍미의
라이트 헤이즐넛
크림**

5. 볼에 달걀흰자를 넣고 휘핑한다.

6. 설탕을 3~4회 나누어 넣어가며 계속해서 휘핑한다.

7. 거품기로 떠 올렸을 때 머랭 뿔이 탄력이 있으면서도 부드럽게 휘어지는 상태가 될 때까지 휘핑한다.

8. 볼에 달걀전란, 헤이즐넛파우더, 슈거파우더, 오렌지제스트, 흑임자 페이스트를 넣고 휘핑한다.

9. 체 친 전분을 넣고 혼합한다.

10. 55~60℃로 녹인 버터를 넣고 혼합한다.

11. 머랭을 2회 나누어 혼합한다.

12. 세사미 스트로이젤에 완성한 흑임자 풍미의 라이트 헤이즐넛 크림을 40g씩 파이핑한 후
    160℃ 오븐에서 15분간 굽는다.

**LIGHT HAZELNUT
CREAM WITH
BLACK SESAME
AROMA**

5. Whip egg whites in a mixing bowl.

6. Gradually add sugar in 3~4 times, and continue whipping.

7. Whip until the meringue forms elastic peaks and bends slightly when the whisk is lifted.

8. In a bowl, whip whole eggs, hazelnut powder, powdered sugar, orange zest, and black sesame paste.

9. Mix with sifted starch.

10. Combine butter melted to 55~60℃.

11. Fold in the meringue in 2 times.

12. Pipe 40g of finished light hazelnut cream with black sesame aroma on the sesame streusel, and bake for 15 minutes at 160℃.

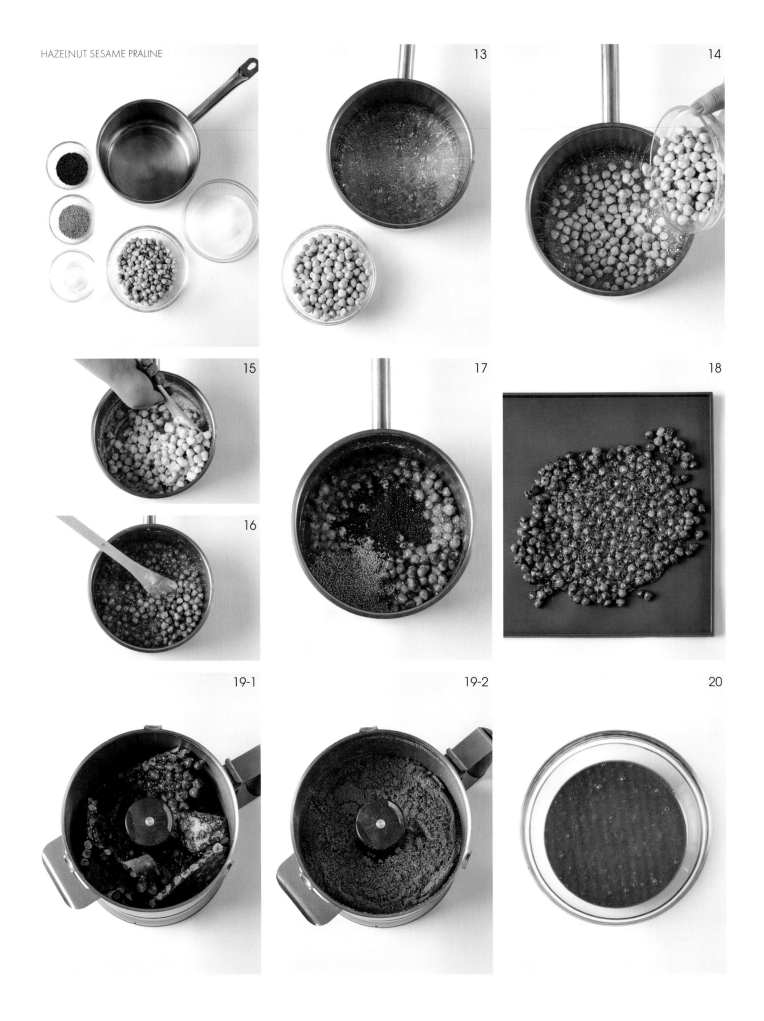

헤이즐넛 세사미 프랄리네

13. 헤이즐넛은 140℃ 오븐에서 약 8분간 구워준다. 냄비에 물과 설탕을 넣고 118~121℃까지 끓인다.

14. 118~121℃가 되면 불에서 내린 후 따뜻한 상태의 헤이즐넛을 넣고 섞어준다.

15. 시럽이 하얗게 재결정화 상태가 될 때까지 계속해서 저어준다.

16. 다시 불에 올려 골고루 저어주며 계속해서 가열한다.

17. 캐러멜화가 시작되면 흑임자와 참깨를 넣고 진한 갈색이 날 때까지 조금 더 진행한다.

18. 실팻에 넓게 펼쳐 완전히 식힌다.

19. 푸드프로세서에 식힌 캐러멜라이즈 헤이즐넛 세사미와 게랑드소금을 넣고 갈아준다.

20. 흐르는 정도의 페이스트 상태가 되면 마무리한다.

HAZELNUT SESAME PRALINE

13. Roast hazelnuts for 8 minutes in an oven preheated to 140℃. In a saucepan, boil water and sugar until 118~121℃.

14. When the temperature reaches 118~121℃, remove from heat and stir in the warm hazelnuts.

15. Continue to stir until the sugar recrystallizes and turns white.

16. Place it back over the heat, and continue to stir.

17. When it starts to caramelize, add black sesame seeds and sesame seeds and continue to cook until the color darkens a bit more.

18. Spread widely over a Silpat, and cool completely.

19. Grind cooled caramelized hazelnuts-sesame and Guérande salt in a food processor.

20. Finish when the texture turns into a fluid paste.

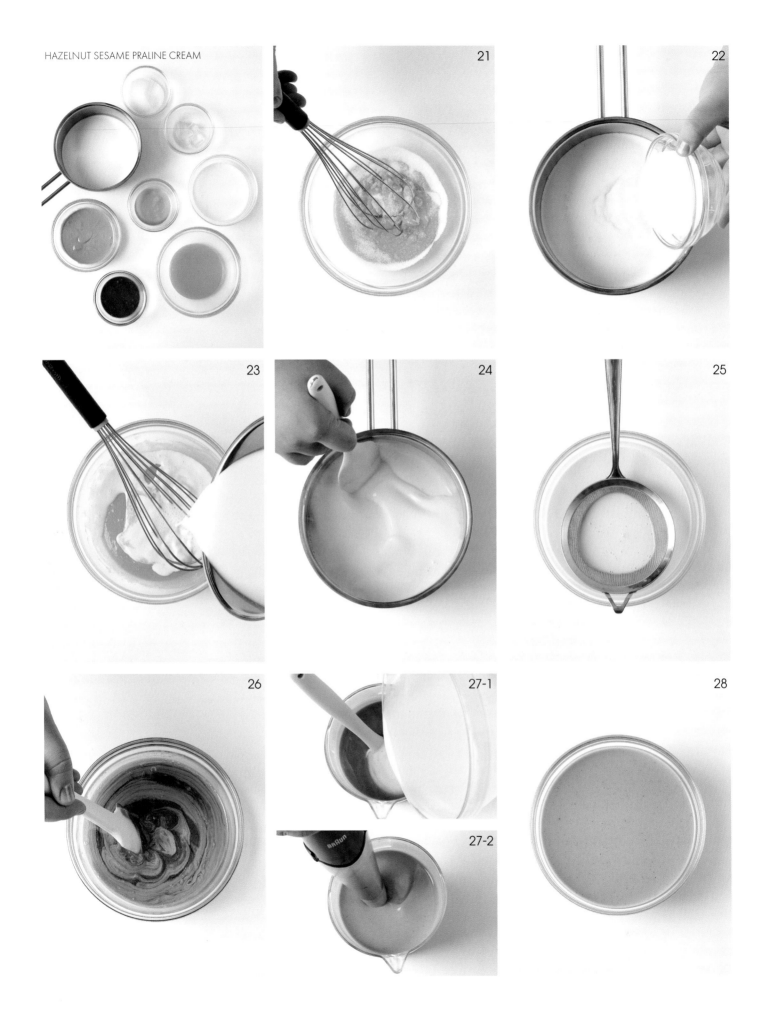

**헤이즐넛
세사미
프랄리네 크림**

21. 달걀노른자와 설탕 1/2을 혼합한다.

22. 냄비에 생크림, 우유, 나머지 설탕을 넣고 45℃로 가열한다.

23. 21에 22를 조금씩 부어가며 혼합한다.

24. 다시 냄비에 옮겨 83~85℃로 가열해 크렘 앙글레이즈를 만든다.

25. 젤라틴매스를 넣고 혼합한 후 체에 거른다. 45℃까지 식힌다.

26. 35℃로 녹인 블론드초콜릿과 카카오버터에 헤이즐넛 세사미 프랄리네를 넣고 혼합한다.

27. 26에 25를 3~4회 나누어 섞고 핸드블렌더로 혼합한다.

28. 냉장고에서 12시간 휴지시킨 후 사용한다.

**HAZELNUT
SESAME
PRALINE CREAM**

21. Mix egg yolks and 1/2 of sugar.

22. In a saucepan, heat fresh cream, milk, and remaining sugar to 45℃.

23. Gradually stir in 22 into 21.

24. Pour back into a saucepan and heat until 83~85℃ to make creme anglaise.

25. Stir in gelatin mass and stain. Cool down to 45℃.

26. Add hazelnut sesame praline into Blond chocolate melted to 35℃ and cacao butter and mix.

27. Mix 25 into 26 in 3~4 times and mix with a hand blender.

28. Rest in the fridge for 12 hours before use.

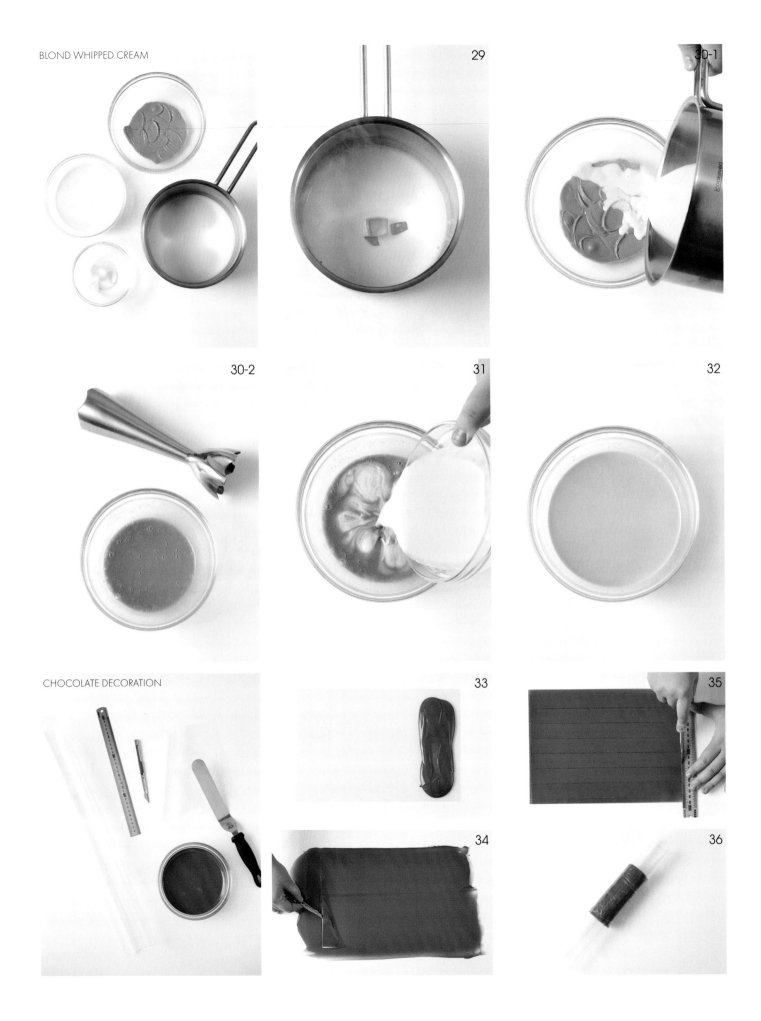

BLOND WHIPPED CREAM

29

30-1

30-2

31

32

CHOCOLATE DECORATION

33

34

35

36

**블론드
휩 크림**

29. 우유를 끓기 직전까지 가열한 후 젤라틴매스를 넣고 혼합한다.

30. 35℃로 녹인 블론드초콜릿에 조금씩 나누어 섞어준 후 핸드블렌더로 혼합한다.

31. 35℃로 데운 생크림을 넣고 계속해서 핸드블렌더로 혼합한다.

32. 냉장고에서 12시간 휴지시킨 후 휘핑해 사용한다.

초콜릿 장식물

33. 투명 필름 위에 템퍼링한 밀크초콜릿을 적당량 부어준다.

34. 스패출러를 이용해 얇게 밀어 편다. 초콜릿이 굳기 시작하면 조심스럽게 필름을 들어 올린다.

35. 폭 2.5cm, 길이 27cm로 커팅한다.

36. 지름 8cm 아크릴 파이프에 감아준다.

**BLOND
WHIPPED
CREAM**

29. Stir in gelatin mass to milk heated until just before boiling.

30. Pour over Blond chocolate melted to 35℃, a little bit at a time, and mix using a hand blender.

31. Add fresh cream warmed to 35℃, and continue to mix with a hand blender.

32. Rest in the fridge for 12 hours. Whip to use.

**CHOCOLATE
DECORATION**

33. Pour a moderate amount of tempered milk chocolate on top of a clear plastic film.

34. Spread thinly using a spatula. When the chocolate starts to crystallize lift the cake collar carefully.

35. Cut into 2.5cm wide, 27cm long strips.

36. Wrap around an 8cm diameter acrylic pipe.

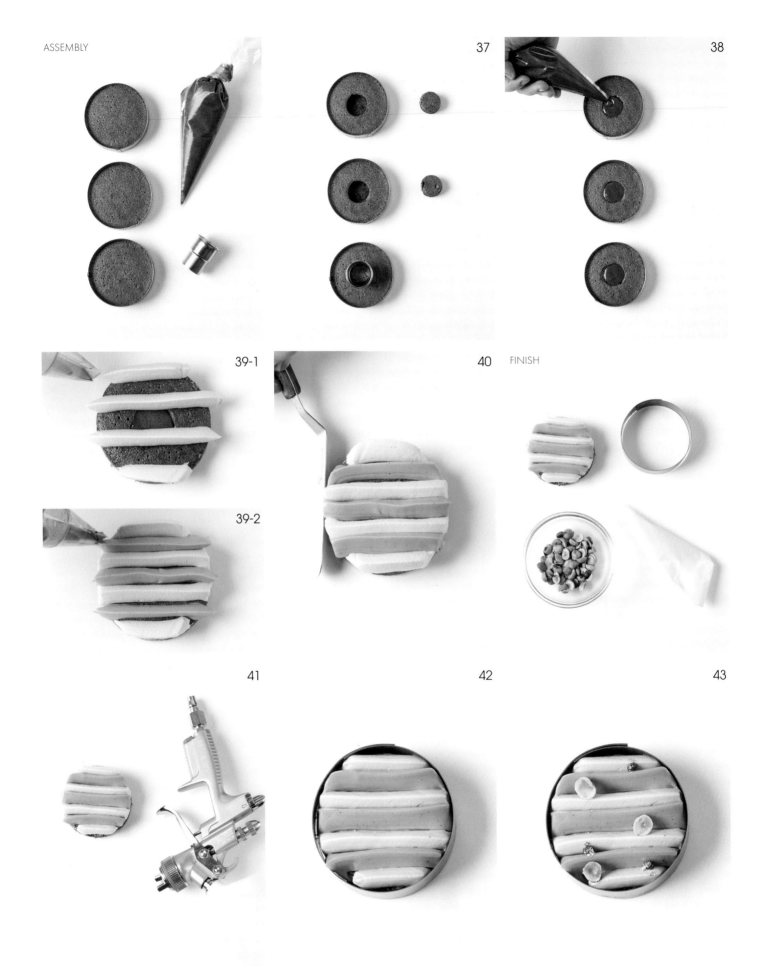

37

38

39-1

39-2

40

41

42

43

조립 **37.** 지름 2.5cm 원형 커터를 이용해 완전히 식힌 흑임자 풍미의 라이트 헤이즐넛 크림 중심부에 구멍을 뚫어 공간을 만든다. 이때 가장 아래쪽에 있는 세사미 스트로이젤까지 뚫지 않도록 주의한다.

38. 만든 공간 안에 헤이즐넛 세사미 프랄리네를 가득 채운 후 냉동고에서 단단하게 굳힌다.

39. 생토노레(15번) 깍지를 이용해 블론드 휩 크림과 세사미 프랄리네 크림을 번갈아가며 파이핑한다.

40. 스패출러로 가장자리를 정리한 후 냉동고에서 단단하게 굳힌다.

마무리 **41.** 크림 표면에 크리스탈 스프레이 건 믹스처(73p)를 분사한다.

42. 높이 2.5cm 띠 모양 초콜릿 장식을 둘러준다.

43. 캐러멜 소스(265p), 헤이즐넛, 식용금박으로 장식해 마무리한다.

ASSEMBLY **37.** In the center of completely cooled light hazelnut cream with black sesame aroma, make a hole using a 2.5cm diameter round cutter. Be careful not to cut through the sesame streusel on the very bottom.

38. Fill the hole with hazelnut sesame praline. Freeze completely.

39. Pipe Blond whipped cream and sesame praline cream alternately using St. Honore nozzles (#15).

40. Organize the edge with a spatula, then freeze completely.

FINISH **41.** Spray crystal spray gun mixture (73p) on the surface of the creams.

42. Wrap with 2.5cm high chocolate strip decoration.

43. Decorate with caramel sauce (265p), hazelnuts and edible gold leaves to finish.

Aroma

COFFEE & HAZELNUT TARTE

EXOTIC CARAMEL TARTE

MATCHA SOUFFLE TARTE

¹² COFFEE & HAZELNUT TARTE

커피 & 헤이즐넛 타르트

ingredients - ∅ 7cm 12ea

파트 아 사블레(41p) or
스위트 반죽 – 글루텐 프리
(53p)

GLUTEN
FREE

장식물

원두 분말

헤이즐넛

식용금박

**커피 풍미의
라이트 헤이즐넛 크림**

헤이즐넛파우더 90g

슈거파우더 70g

달걀전란 80g

인스턴트 커피 4g

커피익스트랙 12g

전분 13g

버터 15g

달걀흰자 24g

설탕 24g

커피 휩 크림

젤라틴매스 15g

우유 105g

원두 12g

블론드초콜릿 119g

(🍫 DULCEY 32%)

생크림 274g

헤이즐넛 커피 프랄리네

헤이즐넛 250g

원두 28g

설탕 250g

물 58g

게랑드소금 2.5g

PATE A SABLÉ(41p) or
**SWEET DOUGH - GLUTEN
FREE**(53p)

GLUTEN
FREE

DECORATION

Ground coffee beans

Hazelnuts

Edible gold leaves

**LIGHT HAZELNUT CREAM
WITH COFFEE AROMA**

90g Hazelnut powder

70g Powdered sugar

80g Whole eggs

4g Instant coffee

12g Coffee extract

13g Starch

15g Butter

24g Egg whites

24g Sugar

COFFEE WHIPPED CREAM

15g Gelatin mass

105g Milk

12g Coffee beans

119g Blond chocolate

(🍫 DULCEY 32%)

274g Fresh Cream

**HAZELNUT COFFEE
PRALINE**

250g Hazelnuts

28g Coffee beans

250g Sugar

58g Water

2.5g Guérande salt

LIGHT HAZELNUT CREAM WITH
COFFEE AROMA

Process

**커피 풍미의
라이트 헤이즐넛
크림**

1. 볼에 달걀흰자를 넣고 휘핑한다.

2. 설탕을 3~4회 나누어 넣어가며 계속해서 휘핑한다.

3. 거품기로 떠 올렸을 때 머랭 뿔이 탄력이 있으면서도 부드럽게 휘어지는 상태가 될 때까지 휘핑한다.

4. 볼에 달걀전란, 슈거파우더, 헤이즐넛파우더, 인스턴트 커피를 넣고 휘핑한다.

5. 커피익스트랙, 전분, 55~60℃로 녹인 버터를 순서대로 넣고 혼합한다.

6. 머랭을 2회 나누어 혼합한다.

7. 80% 소성한 타르트 셸에 완성한 커피 풍미의 라이트 헤이즐넛 크림을 20g씩 채워준다.

8. 헤이즐넛을 적당량 얹고 160℃ 오븐에서 약 15분간 굽는다.

**LIGHT HAZELNUT
CREAM WITH
COFFEE AROMA**

1. Whip egg whites in a mixing bowl.

2. Continue to whip, gradually adding sugar in 3~4 times.

3. Whip until the meringue forms elastic peaks and bends slightly when the whisk is lifted.

4. In a bowl, whip whole eggs, powdered sugar, hazelnut powder, and instant coffee.

5. Mix with coffee extract, starch, and then butter melted to 55~60℃ in order.

6. Fold in the meringue in 2 times.

7. Fill in 20g of the finished light hazelnut cream with coffee aroma in 80% baked tarte shells.

8. Top with a moderate amount of hazelnuts, and bake for 15 minutes at 160℃.

커피 휩 크림

9. 끓기 직전까지 가열한 우유에 160℃ 오븐에서 5분 동안 데운 원두를 넣는다.

10. 핸드블렌더를 이용해 원두를 잘게 분쇄한 후 커피 향을 우린다.

11. 원두를 체로 걸러 제거한다.

12. 젤라틴매스를 넣고 혼합한다.

13. 35℃로 녹인 블론드초콜릿에 조금씩 나누어 섞어준 후 핸드블렌더로 혼합한다.

14. 35℃로 데운 생크림을 넣고 계속해서 핸드블렌더로 혼합한다.

15. 냉장고에서 12시간 휴지시킨 후 휘핑해 사용한다.

COFFEE WHIPPED CREAM

9. Warm coffee beans for 5 minutes in an oven preheated to 160℃, and soak the beans in milk heated until just before boiling.

10. Grind the beans with a hand blender and let them infuse.

11. Remove the beans with a sieve.

12. Stir in gelatin mass.

13. Gradually stir into Blond chocolate melted to 35℃, and mix with a hand blender.

14. Add fresh cream warmed to 35℃, and continue to mix with a hand blender.

15. Rest in the fridge for 12 hours. Whip to use.

**헤이즐넛 커피
프랄리네**

16. 헤이즐넛은 140℃ 오븐에서 약 8분간 구워준다. 냄비에 물과 설탕을 넣고 118~121℃까지 끓인다.

17. 118~121℃가 되면 불에서 내린 후 따뜻한 상태의 헤이즐넛을 넣고 섞어준다.

18. 시럽이 하얗게 재결정화 상태가 될 때까지 계속해서 저어준다.

19. 다시 불에 올려 골고루 저어주며 계속해서 가열한다.

20. 캐러멜화가 시작되면 원두를 넣고 진한 갈색이 날 때까지 조금 더 진행한다.

21. 실팻에 넓게 펼쳐 완전히 식힌다.

22. 푸드프로세서에 식힌 캐러멜라이즈 헤이즐넛 커피와 게랑드소금을 넣고 갈아준다.

23. 흐르는 정도의 페이스트 상태가 되면 마무리한다.

**HAZELNUT
COFFEE
PRALINE**

16. Roast hazelnuts for about 8 minutes in an oven preheated to 140℃. In a saucepan, boil water and sugar until 118~121℃.

17. When the temperature reaches 118~121℃, remove from heat and stir in warmed hazelnuts.

18. Continue to stir until the sugar recrystallizes and turns white.

19. Place it back over the heat, and continue to stir.

20. When it starts to caramelize, add coffee beans and continue to cook until the color darkens a bit more.

21. Spread over a Silpat, and cool completely.

22. Grind cooled caramelized hazelnut coffee and Guérande salt in a food processor.

23. Finish when the texture turns into a fluid paste.

24

25

26

27

28

29

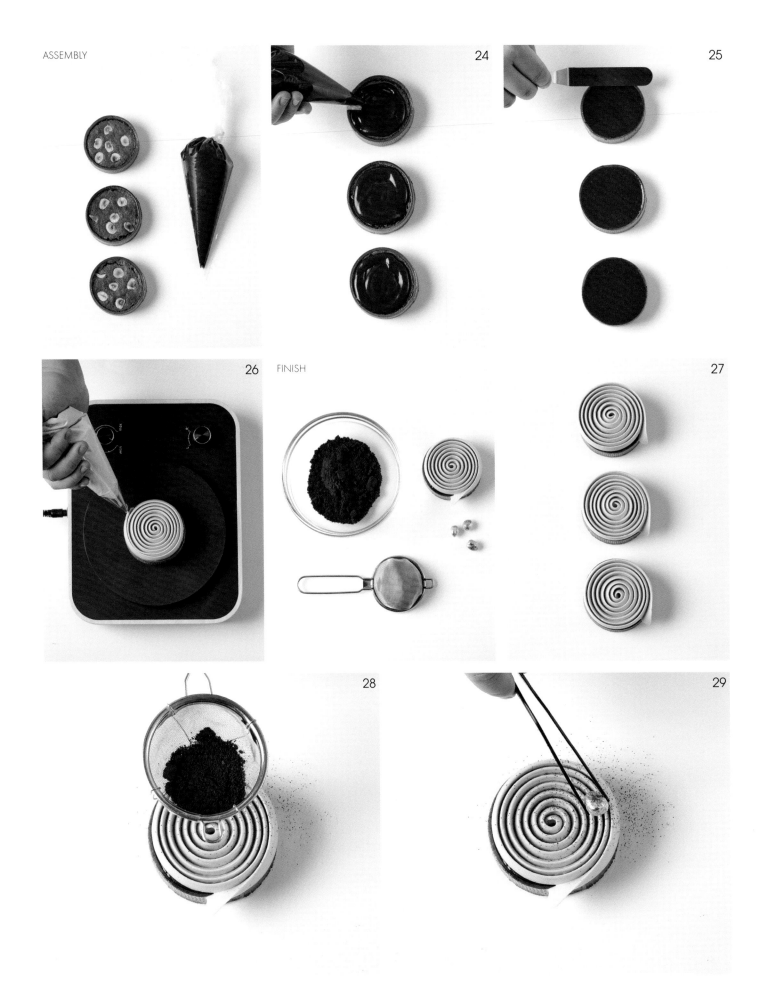

조립	24.	커피 풍미의 라이트 헤이즐넛 크림 위에 헤이즐넛 커피 프랄리네를 가득 채운다.
	25.	스패츌러로 표면을 고르게 정리한 후 냉동고에 넣어 단단하게 굳힌다.
	26.	꽃잎 모양(1303번) 깍지를 이용해 커피 휩 크림을 파이핑한다.
마무리	27.	커피 휩 크림 표면에 크리스탈 스프레이 건 믹스처(73p)를 분사한다.
	28.	체를 이용해 원두 분말을 소량 뿌려준다.
	29.	식용금박을 입힌 헤이즐넛을 올려 마무리한다.

ASSEMBLY	24.	Fill with hazelnut coffee praline on top of the light hazelnut cream with coffee aroma.
	25.	Even out the surface with a spatula and freeze completely.
	26.	Pipe coffee whipped cream with a petal tip (#1303).
FINISH	27.	Spray crystal spray gun mixture (73p) on the surface of the coffee cream.
	28.	Use sieve to sprinkle few of ground coffee beans.
	29.	Place hazelnuts coated with edible gold leaves to finish.

¹³ EXOTIC CARAMEL TARTE

이그조틱 캐러멜 타르트

ingredients - ∅ 8cm 12ea

코코넛 스트로이젤*

버터 71g
설탕 71g
소금 0.9g
아몬드파우더 44g
코코넛 분말 44g
박력분 71g

코코넛 크런치

코코넛 스트로이젤*
183g
에클라도르(🍃) 206g
블론드초콜릿 45g
(🍃 DULCEY 32%)
카카오버터 67g

이그조틱 캐러멜

설탕 142g
패션후르츠 퓌레 79g
코코넛 퓌레 79g
망고 퓌레 71g
블론드초콜릿 71g
(🍃 DULCEY 32%)
바닐라빈 2/3개
버터 59g

블론드초콜릿 무스

물 102g
젤라틴매스 18g
블론드초콜릿 158g
(🍃 DULCEY 32%)
생크림 237g

장식물

블론드초콜릿
(🍃 DULCEY 32%)
카카오파우더
식용금박

글레이즈

물 75g
설탕 150g
물엿 150g
연유 100g
젤라틴매스 60g
블론드초콜릿 150g
(🍃 DULCEY 32%)

COCONUT STREUSEL*

71g Butter
71g Sugar
0.9g Salt
44g Almond powder
44g Coconut powder
71g Cake flour

COCONUT CRUNCH

183g Coconut
        Streusel*
206g Eclat d'Or(🍃)
45g Blond Chocolate
      (🍃 DULCEY 32%)
67g Cacao butter

EXOTIC CARAMEL

142g Sugar
79g Passion fruits
        purée
79g Coconut purée
71g Mango purée
71g Blond chocolate
      (🍃 DULCEY 32%)
2/3pc Vanilla bean
59g Butter

BLONDE CHOCOLATE MOUSSE

102g Water
18g Gelatin mass
158g Blond chocolate
      (🍃 DULCEY 32%)
237g Fresh cream

DECORATION

Blond chocolate
(🍃 DULCEY 32%)
Cacao powder
Edible gold leaves

GLAZE

75g Water
150g Sugar
150g Corn syrup
100g Condensed milk
60g Gelatin mass
150g Blond chocolate
      (🍃 DULCEY 32%)

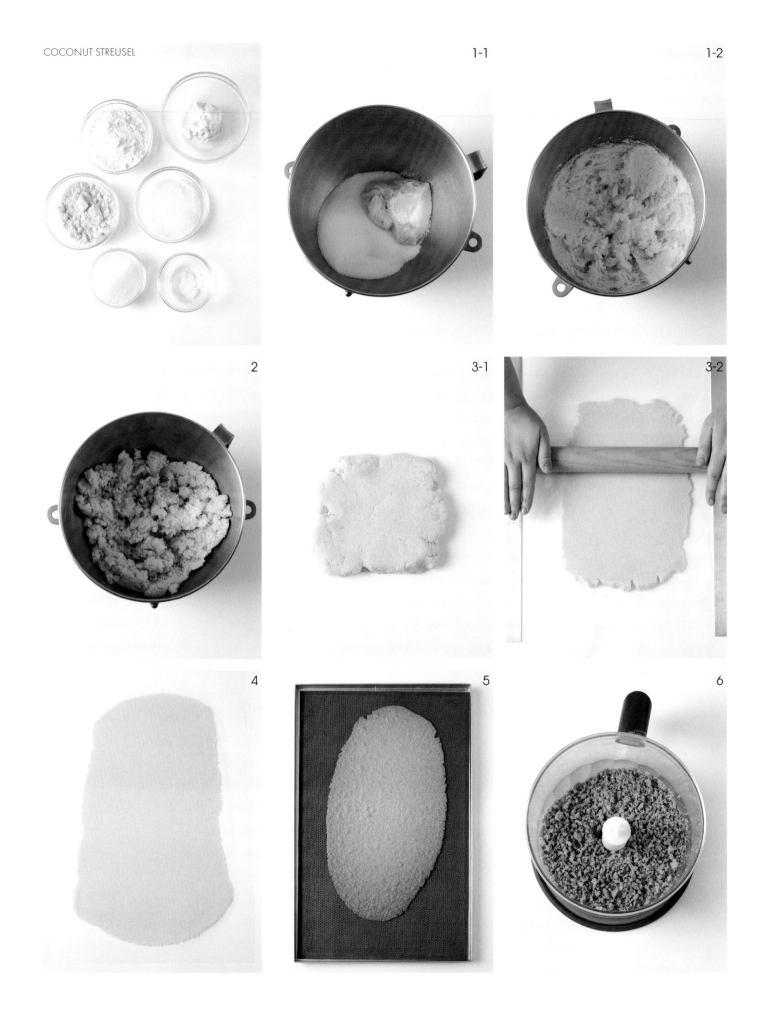

Process

**코코넛
스트로이젤**

1. 버터를 부드럽게 풀어준 후 설탕과 소금을 넣고 혼합한다.

2. 아몬드파우더, 코코넛 분말, 박력분을 넣고 한 덩어리가 될 때까지 믹싱한다.

3. 두 장의 투명 필름 사이에 반죽을 놓고 2mm 두께로 밀어 편다. 이때 2mm 두께의 바를 이용하면
   일정한 두께로 밀어 펼 수 있다.

4. 냉동고에 넣어 단단하게 굳혀준다.

5. 160℃ 오븐에서 15분간 굽는다.

6. 완전히 식힌 후 푸드프로세서를 이용해 갈아준다.

**COCONUT
STREUSEL**

1. Beat butter to soften, and mix with sugar and salt.

2. Mix with almond powder, coconut powder, and cake flour until it forms a ball.

3. Place the dough between two sheets of clear plastic film and roll to 2mm
   thickness. Using 2mm thick bars will help to roll out evenly.

4. Place in the freezer to harden.

5. Bake for 15 minutes at 160℃.

6. Cool completely, then grind in a food processor.

8

9

10

11

코코넛 크런치

7. 코코넛 스트로이젤과 에클라도르를 섞는다.

8. 40℃로 녹인 블론드초콜릿과 카카오버터를 7과 혼합한다.

9. 지름 8cm 원형 틀에 40g씩 분할한다.

10. 스푼을 이용해 가운데가 오목하도록 펼쳐준다.

11. 틀을 제거한 후 냉동고에 넣어 단단하게 굳힌다.

COCONUT CRUNCH

7. Mix coconut streusel and Eclat d'Or.

8. Mix cacao butter and Blond chocolate melted to 40℃ with 7.

9. Divide the mixture 40g each in 8cm diameter ring molds.

10. Use a spoon to spread out, making the center concave.

11. Remove the molds and freeze to harden.

이그조틱
캐러멜

12. 냄비에 패션후르츠 퓌레, 코코넛 퓌레, 망고 퓌레, 바닐라빈을 넣고 끓기 직전까지 가열한다.

13. 다른 냄비에 설탕을 넣고 가열해 캐러멜라이즈한다.

14. **13**에 **12**를 넣고 섞어준다.

15. 체에 걸러준다.

16. 60℃까지 식힌다.

17. 35℃로 녹인 블론드초콜릿에 붓고 핸드블렌더로 혼합한다.

18. 상온 상태의 버터를 넣고 계속해서 핸드블렌더로 혼합한다.

19. 냉장고에서 12시간 휴지시킨 후 사용한다.

EXOTIC
CARAMEL

12. In a saucepan, heat passion fruits purée, coconut purée, mango purée, and vanilla bean until just before boiling.

13. In a separate saucepan, heat sugar to caramelize.

14. Mix **12** into **13**.

15. Strain.

16. Cool down to 60℃.

17. Pour over Blond chocolate melted to 35℃, and mix with a hand blender.

18. Add room temperature butter and continue to mix using a hand blender.

19. Rest in the fridge for 12 hours before use.

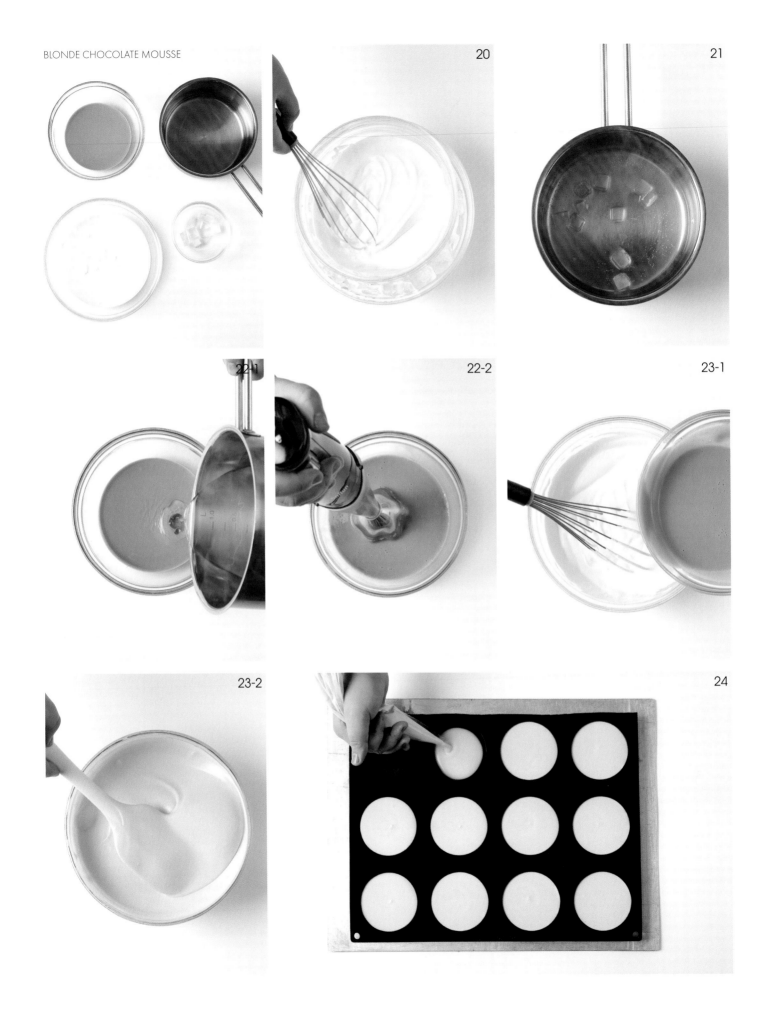

**블론드초콜릿
무스**

20. 생크림은 무스 상태로 휘핑한다.

21. 냄비에 물을 넣고 끓기 직전까지 가열한 후 젤라틴매스를 혼합한다.

22. 35℃로 녹인 블론드초콜릿에 **21**을 붓고 핸드블렌더로 혼합해 가나슈 베이스를 만든다.

23. 무스 상태로 휘핑한 생크림에 혼합한다.

24. 지름 7.5cm, 높이 1.5cm 실리콘 몰드에 완성한 블론드초콜릿 무스를 가득 채운 후 냉장고에서 단단하게 굳힌다.

**BLONDE
CHOCOLATE
MOUSSE**

20. Whip fresh cream to a mousse consistency.

21. Heat water in a saucepan until just before boiling. Stir in gelatin mass.

22. Pour **21** over Blond chocolate melted to 35℃, and mix using a hand blender to make ganache base.

23. Combine with fresh cream whipped to a mousse consistency.

24. In a 7.5cm diameter, 1.5cm high silicon mold, completely fill with finished blond chocolate mousse. Freeze completely.

글레이즈　　25.　냄비에 물과 설탕을 넣고 가열한다.

　　　　　　　26.　시럽 상태가 되면 물엿을 넣고 103℃까지 가열한 후 불에서 내린다.

　　　　　　　27.　연유와 젤라틴매스를 순서대로 혼합한다.

　　　　　　　28.　35℃로 녹인 블론드초콜릿에 붓고 핸드블렌더로 혼합한다.

　　　　　　　29.　냉장고에서 12시간 세팅시킨다.

　　　　　　　30.　사용할 때는 다시 35℃로 온도를 맞춘 다음 핸드블렌더로 균일하게 믹싱해 사용한다.

GLAZE　　25.　In a saucepan, heat water and sugar.

　　　　　　　26.　When it turns into syrup, add corn syrup and boil until 103℃. Remove from heat.

　　　　　　　27.　Stir in condensed milk and gelatin mass in order.

　　　　　　　28.　Pour over Blond chocolate melted to 35℃ and mix using a hand blender.

　　　　　　　29.　Set in the fridge for 12 hours.

　　　　　　　30.　To use, reset the glaze to 35℃, and mix thoroughly using a hand blender.

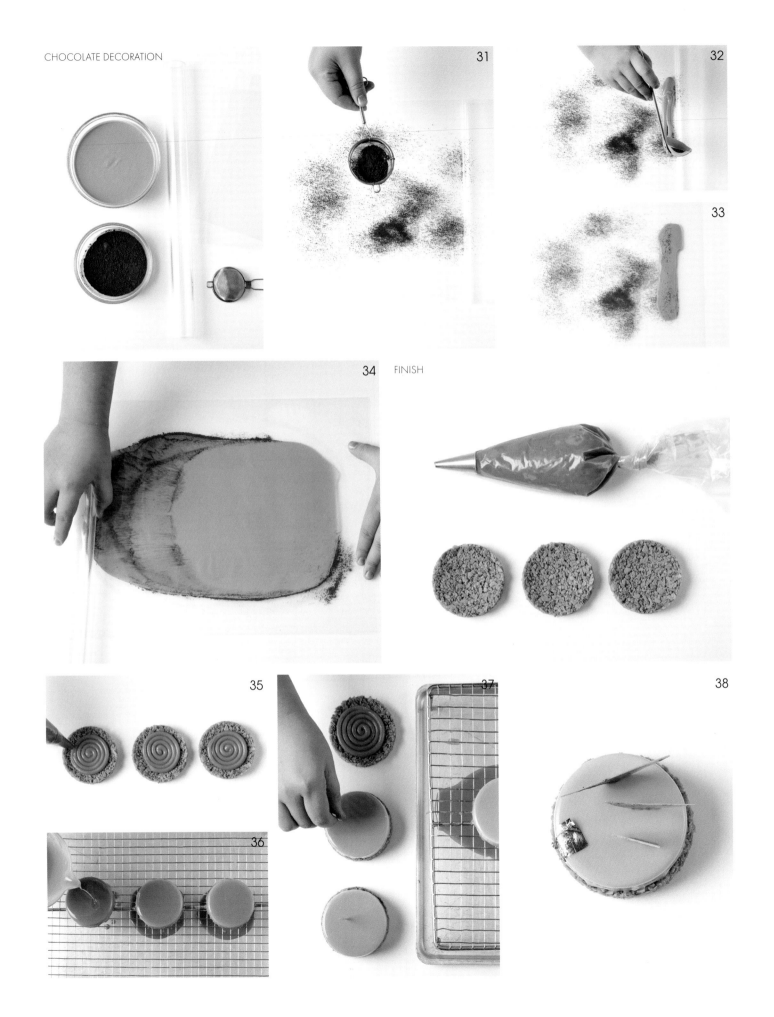

CHOCOLATE DECORATION

31

32

33

34

FINISH

35

36

37

38

초콜릿 장식물

31. 두 장의 투명 필름 사이에 카카오파우더를 뿌려준다.

32. 템퍼링한 블론드초콜릿을 적당량 붓는다.

33. 필름을 덮고 밀대를 이용해 균일한 두께로 밀어 편다.

34. 완전히 굳으면 자연스럽게 부서진 모양을 살려 사용한다.

마무리

35. 코코넛 크런치 위에 이그조틱 캐러멜을 나선형으로 파이핑한다.

36. 단단하게 굳힌 블론드초콜릿 무스를 글레이징한다.

37. 글레이징한 블론드초콜릿 무스를 타르트 위에 올린다.

38. 초콜릿 장식물, 식용금박을 올려 마무리한다.

CHOCOLATE DECORATION

31. Sprinkle cacao powder between two sheets of clear plastic film.

32. Pour a moderate amount of tempered Blond chocolate.

33. Cover with the plastic film and roll out to even thickness using a rolling pin.

34. After the chocolate is completely crystallized, use the naturally broken shapes.

FINISH

35. Pipe exotic caramel in a spiral shape on top of coconut crunch.

36. Glaze completely frozen Blond chocolate mousse.

37. Place glazed Blond chocolate mousse on top of the tarte.

38. Decorate with chocolate decorations and edible gold leaves to finish.

¹⁴ MATCHA SOUFFLE TARTE

말차 수플레 타르트

ingredients - ∅ 7cm 18ea

파트 아 사블레 – 말차(49p)

말차 수플레

크림치즈 160g
설탕A 40g
사워크림 80g
달걀노른자 30g
우유 25g

박력분 20g
말차파우더 10g
버터 33g
달걀흰자 60g
설탕B 30g

기타

당조림 팥
말차파우더

PATE A SABLÉ - MATCHA
(49p)

MATCHA SOUFFLE

160g Cream cheese
40g Sugar A
80g Sour cream
30g Egg yolks
25g Milk

20g Cake flour
10g Matcha powder
33g Butter
60g Egg whites
30g Sugar B

OTHER

Sweet red beans braised
in syrup
Matcha powder

MATCHA SOUFFLE

1

2

3

4

5

6

FINISH

7

8

9

Process

말차 수플레

1. 볼에 달걀흰자를 넣고 휘핑한다.

2. 설탕B를 3~4회 나누어 넣어가며 계속해서 휘핑한다.

3. 거품기로 떠 올렸을 때 머랭 뿔이 부드럽게 휘어지는 상태가 될 때까지 휘핑한다.

4. 푸드프로세서에 크림치즈, 설탕A, 사워크림, 달걀노른자, 우유, 박력분, 말차파우더를 넣고 균일한 상태가 될 때까지 믹싱한다.

5. 55~60℃로 녹인 버터를 혼합한다.

6. 머랭을 2회 나누어 혼합한다.

마무리

7. 80% 소성한 타르트 셸에 당조림 팥을 10g씩 채운다.

8. 말차 수플레 반죽을 22g씩 파이핑한 후 데크 오븐 기준 윗불 170℃, 아랫불 170℃에서 13분간 굽는다.

9. 말차파우더를 뿌려 마무리한다.

MATCHA SOUFFLE

1. Whip egg whites in a mixing bowl.

2. Continue to whip, gradually adding sugar B in 3~4 times.

3. Whip until the meringue forms elastic peaks and bends slightly when the whisk is lifted.

4. In a food processor, add cream cheese, sugar A, sour cream, egg yolks, milk, cake flour, and matcha powder and mix until homogeneous.

5. Mix with butter melted to 55~60℃.

6. Fold in the meringue in 2 times.

FINISH

7. Fill 10g of sweet red beans braised in syrup in 80% baked tarte shells.

8. Pipe 22g of matcha souffle batter in each shell. Bake in deck oven for 13 minutes at top heat 170℃, bottom heat 170℃.

9. Sprinkle matcha powder to finish.

Cream & Milk

CREAM BRULEE & BANANA TARTE

MASCARPONE & BLUEBERRY TARTE

DOUBLE CHEESE STRAWBERRY TARTE

15 CREAM BRULEE & BANANA TARTE

크림 브륄레 & 바나나 타르트

ingredients - Ø 7.5cm 12ea

GLUTEN FREE

아몬드 페이스트*

아몬드 125g
슈거파우더 25g

현미 & 아몬드 크런치

화이트초콜릿 76g
(IVOIRE 35%)
아몬드 페이스트* 67g
소금 1.1g
아몬드 오일 15g
볶은 현미 141g

캐러멜 소스

물엿 99g
설탕 99g
생크림 151g
바닐라빈 1/3개
소금 1.6g
젤라틴매스 4.8g
버터 49g

크림 브륄레

설탕 112g
달걀노른자 112g
달걀흰자 112g
생크림 588g
럼 37g
바닐라빈 1.5개

기타

바나나
카소나드

GLUTEN FREE

ALMOND PASTE*

125g Almonds
25g Powdered
       sugar

BROWN RICE & ALMOND CRUNCH

76g White chocolate
(IVOIRE 35%)
67g Almond paste*
1.1g Salt
15g Almond oil
141g Roasted
       brown rice

CARAMEL SAUCE

99g Corn syrup
99g Sugar
151g Fresh cream
1/3pc Vanilla bean
1.6g Salt
4.8g Gelatin mass
49g Butter

CREAM BRULEE

112g Sugar
112g Egg yolks
112g Egg whites
588g Fresh cream
37g Rum
1.5pc Vanilla beans

OTHER

Bananas
Cassonade

ALMOND PASTE

1

2-1

2-2

2-3

BROWN RICE & ALMOND CRUNCH

3

4

5

Process

**아몬드
페이스트**

1. 아몬드는 140℃ 오븐에서 약 30분간 로스팅한다.

2. 완전히 식힌 아몬드와 슈거파우더를 푸드프로세서에 넣고 페이스트 상태가 될 때까지 갈아준다.

**현미 & 아몬드
크런치**

3. 볶은 현미에 소금, 40℃로 녹인 화이트초콜릿, 아몬드 페이스트, 아몬드 오일을 넣고 혼합한다.

4. 지름 7.5cm 원형 틀에 20g씩 분할한 후 스푼을 이용해 가운데가 오목하도록 펼쳐준다.

5. 틀을 제거한 후 냉동고에서 단단하게 굳혀준다.

**ALMOND
PASTE**

1. Roast almonds for about 30 minutes at 140℃.

2. Grind completely cooled almonds and powdered sugar in a food processor until the texture turns into a paste.

**BROWN RICE
& ALMOND
CRUNCH**

3. Mix roasted brown rice, salt, white chocolate melted to 40℃, almond paste, and almond oil.

4. Divide the mixture 20g each in 7.5cm diameter ring molds. Use a spoon to spread out, making the center concave.

5. Remove the molds and freeze to harden.

캐러멜 소스

6. 생크림에 바닐라빈과 소금을 넣고 끓기 직전까지 가열한다.

7. 냄비에 물엿과 설탕을 넣고 가열해 캐러멜라이즈한다.

8. 7에 6을 붓고 섞어준 후 108℃까지 가열한다.

9. 60℃까지 식힌 후 젤라틴매스를 넣고 섞는다.

10. 상온 상태의 버터를 넣고 핸드블렌더로 혼합한다. 상온에서 12시간 휴지시켜 사용한다.

CARAMEL SAUCE

6. Add vanilla beans and salt to fresh cream and heat until just before boiling.

7. In a separate saucepan, heat corn syrup and sugar to caramelize.

8. Slowly pour 6 into 7. Mix and heat until 108℃.

9. Cool down to 60℃. Stir in gelatin mass.

10. Add room temperature butter and mix with a hand blender. Rest in ambient temperature for 12 hours before use.

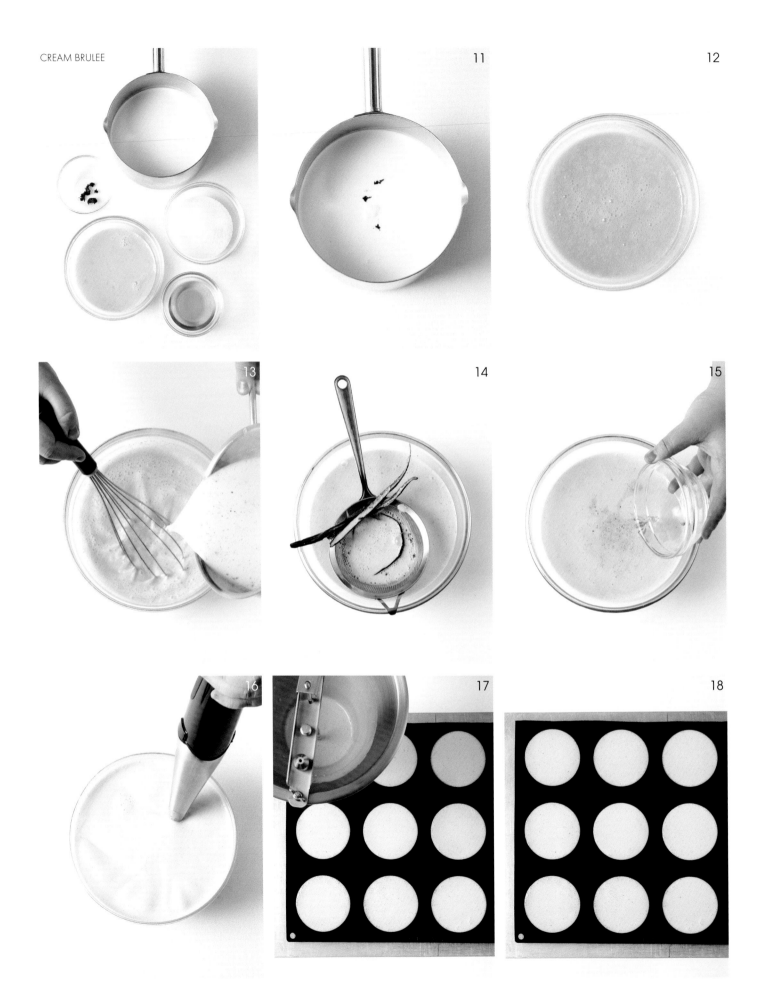

크림 브륄레

11. 냄비에 생크림, 설탕 1/2, 바닐라빈을 넣고 80℃로 데운 후 바닐라 향을 충분히 우린다.

12. 볼에 달걀노른자, 달걀흰자, 나머지 설탕을 넣고 섞는다.

13. 11을 조금씩 부어가며 섞는다.

14. 체에 걸러준다.

15. 럼을 넣고 섞는다.

16. 핸드블렌더로 혼합한 후 냉장고에서 12시간 휴지시킨다.

17. 지름 7.5cm, 높이 1.5cm 원형 실리콘 몰드에 40g씩 채운다.

18. 80℃로 예열된 오븐에서 약 60분간 구워준 후 냉동고에서 단단하게 굳힌다.

CREAM BRULEE

11. In a saucepan, heat fresh cream, 1/2 of sugar, and vanilla bean to 80℃, and sufficiently infuse vanilla aroma.

12. In a bowl, mix egg yolks, egg whites, and remaining sugar.

13. Gradually stir in **11**, a little bit at a time.

14. Strain.

15. Mix with rum.

16. Mix using a hand blender, and rest in the fridge for 12 hours.

17. Fill 40g in each round silicone molds of 7.5cm diameter, 1.5cm high.

18. Bake for 60 minutes in an oven preheated to 80℃. Freeze completely.

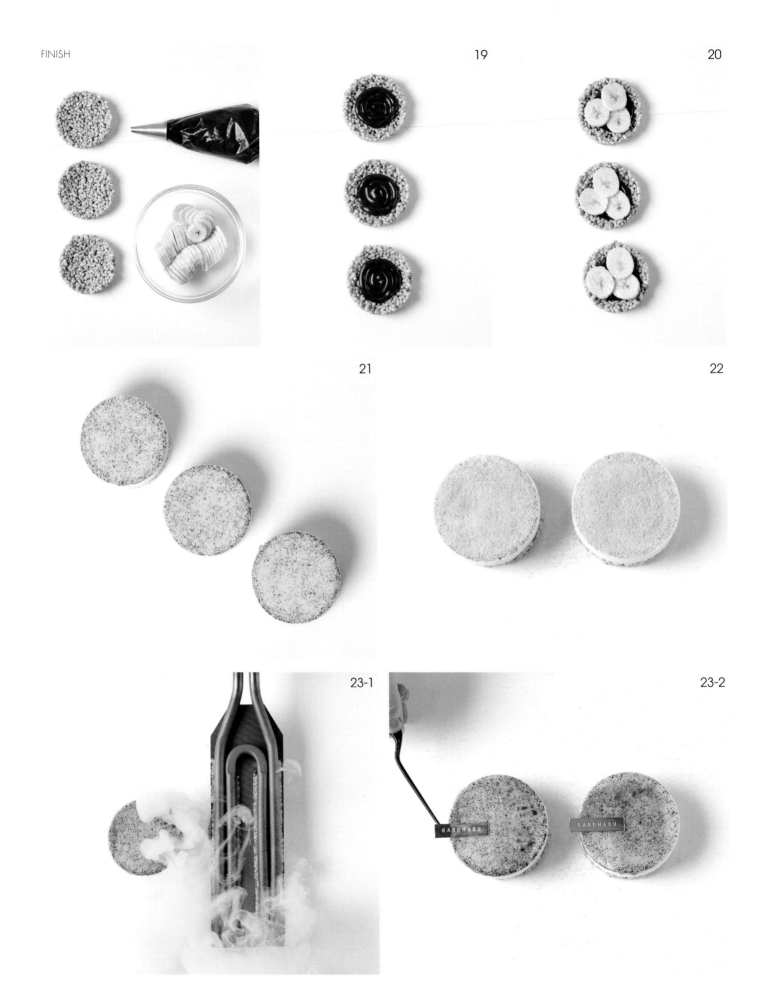

마무리

19. 현미&아몬드 크런치에 캐러멜 소스를 나선형으로 파이핑한다.

20. 슬라이스한 바나나를 3조각씩 올린다.

21. 몰드에서 빼낸 크림 브륄레를 올린다.

22. 크림 브륄레 윗면에 카소나드를 충분히 뿌린다.

23. 카라멜라이저 또는 토치를 이용해 윗면을 캐러멜라이징한다.

FINISH

19. Pipe caramel sauce in a spiral shape on brown rice & almond crunch.

20. Put three pieces of sliced banana on each tartes.

21. Remove cream brulee from the mold and place it on top.

22. Sprinkle a good amount of cassonade on top of the cream brulee.

23. Caramelize the top using a caramelizer or blow torch.

¹⁶ MASCARPONE & BLUEBERRY TARTE

마스카르포네 & 블루베리 타르트

ingredients - Ø 7cm 12ea

파트 아 사블레(41p)
or **스위트 반죽 – 글루
텐 프리**(53p)

GLUTEN
FREE

장식물

레몬 머랭 쿠키*
슈거파우더
민트 잎
식용꽃

**레몬 풍미의
라이트 아몬드 크림**

아몬드파우더 90g
슈거파우더 70g
달걀전란 80g
레몬제스트 레몬 1개
분량
전분 13g
버터 15g
달걀흰자 24g
설탕 24g

블루베리 & 레몬 잼

냉동 야생 블루베리
158g
레몬주스 79g
바닐라빈 1/4개
설탕 60g
NH펙틴 3g

마스카르포네 휩 크림

우유 65g
젤라틴매스 7.2g
화이트초콜릿 72g
(IVOIRE 35%)
생크림 183g
마스카르포네치즈 78g

기타

블루베리

레몬 머랭 쿠키*

달걀흰자 50g
설탕 50g
슈거파우더 50g
레몬제스트 레몬 1/2개
분량

**초콜릿 스프레이 건
믹스처**

카카오버터 90g
화이트초콜릿 90g
(OPALYS 33%)

PATE A SABLÉ(41p)
or **SWEET DOUGH –
GLUTEN FREE**(53p)

GLUTEN
FREE

DECORATION

Lemon meringue
cookies *
Powdered sugar
Mint leaves
Edible flowers

**LIGHT ALMOND
CREAM WITH LEMON
AROMA**

90g Almond powder
70g Powdered sugar
80g Whole eggs
Zest of 1 lemon
13g Starch
15g Butter
24g Egg whites
24g Sugar

**BLUEBERRY & LEMON
JAM**

158g Frozen wild
blueberries
79g Lemon juice
1/4pc Vanilla bean
60g Sugar
3g Pectin NH

**MASCARPONE
WHIPPED CREAM**

65g Milk
7.2g Gelatin mass
72g White chocolate
(IVOIRE 35%)
183g Fresh cream
78g Mascarpone
cheese

OTHER

Blueberries

**LEMON MERINGUE
COOKIES***

50g Egg whites
50g Sugar
50g Powdered sugar
Zest of 1/2 lemon

**CHOCOLATE SPRAY
GUN MIXTURE**

90g Cacao butter
90g White chocolate
(OPALYS 33%)

LIGHT ALMOND CREAM WITH
LEMON AROMA

Process

레몬 풍미의
라이트
아몬드 크림

1. 볼에 달걀흰자를 넣고 휘핑한다.

2. 설탕을 3~4회 나누어 넣어가며 계속해서 휘핑한다.

3. 거품기로 떠 올렸을 때 머랭 뿔이 탄력이 있으면서도 부드럽게 휘어지는 상태가 될 때까지 휘핑한다.

4. 볼에 달걀전란, 아몬드파우더, 슈거파우더, 레몬제스트를 넣고 휘핑한다.

5. 체 친 전분과 55~60℃로 녹인 버터를 순서대로 넣고 혼합한다.

6. 머랭을 2회 나누어 혼합한다.

7. 80% 소성한 타르트 셸에 완성한 레몬 풍미의 라이트 아몬드 크림을 20g씩 채워준다.

8. 반으로 가른 블루베리를 적당량 올린 후 160℃ 오븐에서 약 15분간 굽는다.

LIGHT ALMOND
CREAM WITH
LEMON AROMA

1. Whip egg whites in a mixing bowl.

2. Continue to whip, gradually adding sugar in 3~4 times.

3. Whip until the meringue forms elastic peaks and bends slightly when the whisk is lifted.

4. In a bowl, whip whole eggs, almond powder, powdered sugar, and lemon zest.

5. Mix with sifted starch and butter melted to 55~60℃ in order.

6. Fold in the meringue in 2 times.

7. Fill in 20g of the finished light almond cream with lemon aroma in 80% baked tarte shells.

8. Top with a moderate amount of blueberry halves, and bake for 15 minutes at 160℃.

블루베리 &
레몬 잼

9. 냄비에 냉동 야생 블루베리, 레몬주스, 바닐라빈을 넣고 45℃까지 가열한다.

10. 미리 혼합해둔 설탕과 NH펙틴을 넣는다.

11. 한번 끓어오를 때까지 가열한다.

12. 완전히 식힌 후 지름 3.5cm 반구형 실리콘 몰드에 가득 채워준다.

13. 스패출러로 표면을 고르게 정리한 후 냉동고에 넣어 단단하게 굳힌다.

BLUEBERRY &
LEMON JAM

9. In a saucepan, heat frozen wild blueberries, lemon juice, and vanilla bean until 45℃.

10. Stir in previously mixed sugar and pectin NH.

11. Heat until the mixture boils once.

12. Cool completely, then fill in 3.5cm diameter half-sphere silicone molds.

13. Even out the surface with a spatula. Freeze completely.

**마스카르포네
휩 크림**

14. 우유를 끓기 직전까지 가열한 후 젤라틴매스를 혼합한다.

15. 35℃로 녹인 화이트초콜릿에 **14**를 붓고 핸드블렌더로 혼합한다.

16. 35℃로 데운 생크림, 마스카르포네치즈를 넣는다.

17. 계속해서 핸드블렌더로 혼합한다.

18. 냉장고에서 12시간 휴지시킨 후 휘핑해 사용한다.

**MASCARPONE
WHIPPED
CREAM**

14. Stir in gelatin mass to milk heated until just before boiling.

15. Pour **14** over white chocolate melted to 35℃, and mix with a hand blender.

16. Add fresh cream warmed to 35℃ and mascarpone cheese.

17. Continue to mix using a hand blender.

18. Rest in the fridge for 12 hours. Whip to use.

19

20

21-1

21-2

22

23

CHOCOLATE SPRAY GUN MIXTURE

24

레몬 머랭 쿠키	19.	냄비에 달걀흰자와 설탕을 넣고 65℃까지 가열한다.
	20.	머랭이 하얗게 올라오고 힘 있는 상태가 될 때까지 휘핑한다.
	21.	레몬제스트와 체 친 슈거파우더를 넣고 혼합한다.
	22.	원형(지름 0.8cm) 깍지를 이용해 원뿔 모양으로 파이핑한다.
	23.	70℃ 오븐에서 1시간 동안 건조시킨다.

초콜릿 스프레이 건 믹스처	24.	녹인 카카오버터와 화이트초콜릿을 혼합한다. 완성한 레몬 머랭 쿠키에 분사한다.
		* 초콜릿 스프레이 건 믹스처는 머랭에 수분이 흡수되는 것을 방지해 바삭한 식감을 오랫동안 유지할 수 있게 해준다.

LEMON MERINGUE COOKIES	19.	In a saucepan, heat egg whites and sugar until 65℃.
	20.	Whip until the meringue turns white and obtains a firm structure.
	21.	Mix with lemon zest and sifted powdered sugar.
	22.	Pipe in cone shapes using a round tip (0.8cm diameter).
	23.	Dry for 1 hour at 70℃.

CHOCOLATE SPRAY GUN MIXTURE	24.	Mix melted cacao butter and white chocolate. Spray on finished lemon meringue cookies.
		* Chocolate spray gun mixture prevents moisture from being absorbed into the meringue, allowing it to maintain its crispy texture longer.

25

26

27

28

29

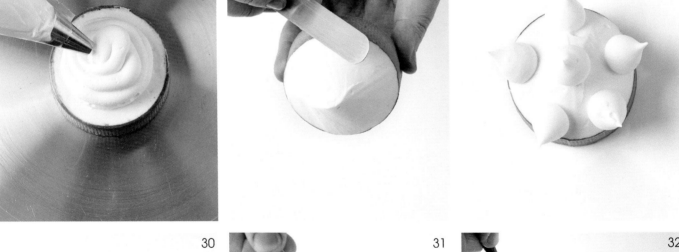

30

31

32

| 조립 | 25. | 레몬 풍미의 라이트 아몬드 크림 위에 마스카르포네 휩 크림을 가득 채운 후 표면을 고르게 정리한다. |
| | 26. | 실리콘 몰드에서 빼낸 블루베리&레몬 잼을 올려준다. |

마무리	27.	마스카르포네 크림을 모양내어 파이핑한다.
	28.	스패츌러를 이용해 크림 표면을 삼각뿔 모양으로 매끄럽게 정리한다.
	29.	레몬 머랭 쿠키를 올린다.
	30.	블루베리를 올린다.
	31.	슈거파우더를 뿌린다.
	32.	민트 잎, 식용꽃을 올려 마무리한다.

| ASSEMBLY | 25. | Fill with mascarpone whipped cream on top of the light almond cream with lemon aroma, and even out the surface. |
| | 26. | Remove blueberry & lemon jam from the mold, and place it on top of the cream. |

FINISH	27.	Pipe a heapful amount of mascarpone whipped cream.
	28.	Use a spatula to shape the cream into a cone.
	29.	Arrange lemon meringue cookies.
	30.	Place blueberries in between.
	31.	Dust with powdered sugar.
	32.	Decorate with mint leaves and edible flowers to finish.

¹⁷ DOUBLE CHEESE STRAWBERRY TARTE

더블 치즈 스트로베리 타르트

ingredients - ⌀ 7cm 12ea

파트 아 사블레(41p) or **스위트 반죽 – 글루텐 프리**(53p)

GLUTEN FREE

라임 치즈케이크

크림치즈 269g
설탕 51g
달걀전란 96g
생크림 214g
라임주스 19g
라임제스트 라임 1개 분량

딸기 젤리

냉동 야생 딸기 77g
딸기 퓌레 118g
설탕 15g
전분 7g
젤라틴매스 6g

치즈 크림

크림치즈 134g
마스카르포네치즈 37g
연유 32g
설탕 32g
생크림 228g
레몬주스 26g
젤라틴매스 11.8g

초콜릿 스프레이 건 믹스처(281p)

카카오버터 90g
화이트초콜릿 90g
(OPALYS 33%)

장식물

딸기
민트 잎
라임제스트
코코넛 분말

PATE A SABLÉ(41p) or **SWEET DOUGH - GLUTEN FREE**(53p)

GLUTEN FREE

LIME CHEESECAKE

269g Cream cheese
51g Sugar
96g Whole eggs
214g Fresh cream
19g Lime juice
Zest of 1 lime

STRAWBERRY JELLY

77g Frozen wild strawberries
118g Strawberry purée
15g Sugar
7g Starch
6g Gelatin mass

CHEESE CREAM

134g Cream cheese
37g Mascarpone cheese
32g Condensed milk
32g Sugar
228g Fresh cream
26g Lemon juice
11.8g Gelatin mass

CHOCOLATE SPRAY GUN MIXTURE(281P)

90g Cacao butter
90g White chocolate
(OPALYS 33%)

DECORATION

Strawberries
Mint leaves
Lime zest
Coconut powder

Process

라임 치즈케이크		
	1.	푸드프로세서에 크림치즈와 설탕을 넣고 부드럽게 풀어준다.
	2.	달걀전란과 생크림을 순서대로 넣어가며 계속해서 혼합한다.
	3.	라임제스트와 라임주스를 넣고 혼합한다.
	4.	80% 소성한 타르트 셸에 완성한 라임 치즈케이크를 50g씩 채워준 후 데크 오븐 기준 윗불 160℃, 아랫불 140℃에서 약 30분간 굽는다.

LIME CHEESECAKE		
	1.	Using a food processor, soften cream cheese and sugar together.
	2.	Continue to mix with whole eggs and fresh cream in order.
	3.	Blend with lime zest and lime juice.
	4.	Fill 50g of the finished lime cheesecake batter in 80% baked tarte shells. Bake for about 30 minutes in a deck oven, at top heat 160℃, bottom heat 140℃.

딸기 젤리

5. 냄비에 냉동 야생 딸기와 딸기 퓌레를 넣고 45℃로 가열한다.

6. 미리 혼합해둔 설탕과 전분을 넣고 전분이 충분히 호화될 때까지 가열한다.

7. 불에서 내린 후 젤라틴매스를 혼합한다.

8. 핸드블렌더를 이용해 균일하게 믹싱한다.

9. 지름 3.5cm 반구형 실리콘 몰드에 채운다.

10. 냉동고에서 단단하게 굳힌다.

STRAWBERRY JELLY

5. In a saucepan, heat frozen wild strawberries and strawberry purée until 45℃.

6. Stir in previously mixed sugar and starch, and continue to cook until the starch is gelatinized enough.

7. Remove from heat, and stir in gelatin mass.

8. Mix evenly with a hand blender.

9. Fill in 3.5cm diameter half-sphere silicone molds.

10. Freeze completely.

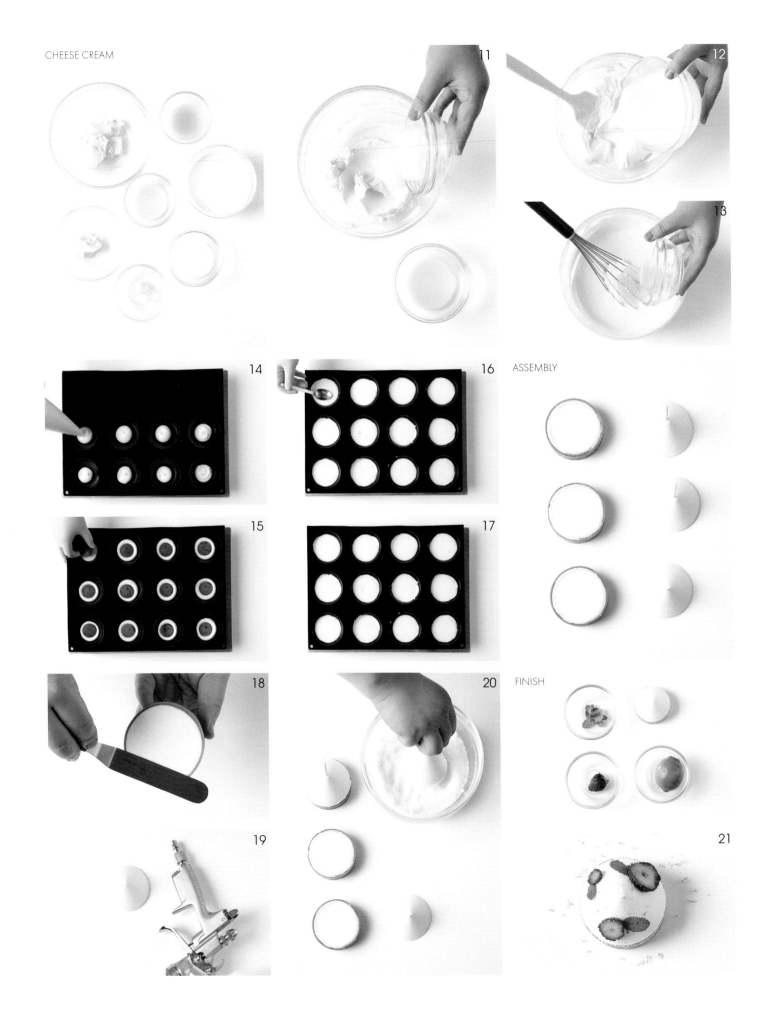

CHEESE CREAM

11

12

13

14

15

16

17

ASSEMBLY

18

19

20

FINISH

21

치즈 크림	11.	크림치즈와 마스카르포네 치즈를 부드럽게 풀어준 후 연유와 설탕을 넣고 섞는다.
	12.	치즈가 덩어리지지 않도록 생크림을 천천히 나누어가며 혼합한다.
	13.	레몬주스와 45℃로 녹인 젤라틴매스를 순서대로 넣고 섞는다.
	14.	가볍게 휘핑한 후 지름 8cm 원뿔 모양 실리콘 몰드에 50% 채운다.
	15.	냉동고에서 굳힌 딸기 젤리를 넣는다.
	16.	다시 치즈 크림을 채우고 스푼을 이용해 표면을 고르게 정리한다.
	17.	냉동고에서 단단하게 굳힌다. 남은 크림은 마무리 작업을 위해 냉장 보관해둔다.

조립	18.	라임 치즈케이크 위 빈 공간에 남은 치즈 크림을 가득 채운 후 표면을 고르게 정리한다.
	19.	몰드에서 빼낸 치즈 크림 표면에 초콜릿 스프레이 건 믹스처(281p), 크리스탈 스프레이 건 믹스처(73p) 순서로 분사한다.
	20.	치즈 크림 가장자리에 코코넛 분말을 장식한 후 타르트 셸 위에 올린다.

| 마무리 | 21. | 라임제스트, 슬라이스한 딸기, 민트 잎으로 장식해 마무리한다. |

CHEESE CREAM	11.	Soften cream cheese and mascarpone cheese, then combine with condensed milk and sugar.
	12.	Gradually mix with fresh cream, taking care to incorporate without any lumps.
	13.	Mix with lemon juice and gelatin mass melted to 45℃ in order.
	14.	Lightly whip and fill 50% of 8cm diameter cone-shaped silicone molds.
	15.	Insert the frozen strawberry jelly.
	16.	Fill with the cheese cream again on top, then even out the surface with a spoon.
	17.	Freeze completely. Store the remaining cream in the fridge for the finishing touch.

ASSEMBLY	18.	Fill with the remaining cheese cream in the hollow space of the lime cheesecake.
	19.	Remove the cheese cream from the molds. Spray the surface with chocolate spray gun mixture (281p), then with crystal spray gun mixture (73p) in order.
	20.	Decorate the edge of the cheese cream with coconut powder, and place it on the tarte.

| FINISH | 21. | Decorate with lime zest, sliced strawberries, and mint leaves to finish. |

Savory

BACON & POTATO QUICHE

SEAFOOD & TOMATO QUICHE

¹⁸ BACON & POTATO QUICHE

베이컨 & 감자 키슈

ingredients - ⌀ 16cm 2ea

파트 아 브리제(57p) or
파이 반죽 – 글루텐 프리
(61p)

GLUTEN FREE   NUTS FREE

아파레유

생크림 125g

우유 63g

달걀 75g

소금 2g

후추 적당량

가르니튀르

베이컨	크림치즈
양파	체다치즈
마늘	고다치즈
감자	방울토마토
브로콜리	로즈마리
포르치니 버섯	소금
청양고추	후추

장식물

트러플

기타

올리브오일

PATE A BRISÉE(57p) or
**PIE DOUGH - GLUTEN
FREE**(61p)

GLUTEN FREE   NUTS FREE

APPAREIL

125g Fresh cream

63g Milk

75g Eggs

2g Salt

QS Ground pepper

GARNITURE

Bacon	Cream cheese
Onion	Cheddar cheese
Garlic	Gouda cheese
Potatoes	Cherry tomatoes
Broccoli	Rosemary
Porcini Mushrooms	Salt
Cheongyang pepper (or other spicy chili pepper)	Ground pepper

DECORATION

Truffle

OTHER

Olive oil

APPAREIL

1

GARNITURE

2

3

4

5

FINISH

6

7

8

Process

아파레유   1.   볼에 모든 재료를 넣고 혼합한다

가르니튀르   2.   마늘과 양파는 올리브오일을 두른 팬에 볶는다. 감자는 사각으로 썰어 끓는 물에 1분간 데친다. 브로콜리는 깨끗하게 세척한 후 끓는 물에 데친다. 포르치니 버섯은 물에 불린 후 올리브오일에 볶는다. 크림치즈는 사방 1cm 크기로, 베이컨과 청양고추는 적당한 크기로 자른다.

   3.   방울토마토를 반으로 가른 후 팬에 고르게 펼쳐준다.

   4.   올리브오일, 로즈마리, 소금, 후추를 적당량 뿌려준다.

   5.   골고루 버무린 다음 200℃ 오븐에서 10분간 굽다가 150℃로 낮춰 15분간 굽는다.

마무리   6.   80% 소성한 셸 안에 준비한 가르니튀르를 채우고 체다치즈와 고다치즈를 적당량 올린다.

   7.   아파레유를 가득 붓고 175℃ 오븐에서 약 40분간 굽는다.

   8.   올리브오일을 적당량 두르고 슬라이스 트러플을 얹어 완성한다.

APPAREIL   1.   Mix all ingredients in a bowl.

GARNITURE   2.   Stir fry garlic and onion with olive oil. Dice potatoes and blanch in boiling water for 1 minute. Wash broccoli thoroughly and blanch in boiling water. If porcini mushrooms are dried, soak in water, then stir fry with olive oil. Dice cream cheese in 1cm cubes. Chop bacon and Cheongyang pepper to moderate size.

   3.   Halve the cherry tomatoes and spread out evenly on a baking tray.

   4.   Drizzle olive oil and sprinkle rosemary leaves, salt, and pepper.

   5.   Toss to coat evenly. Bake for 10 minutes at 200℃, then reduce heat to 150℃ and continue to bake for 15 minutes.

FINISH   6.   In 80% baked tarte shells, fill with the garnitures and top with a moderate amount of cheddar and gouda.

   7.   Top it off with the appareil, and bake for 40 minutes at 175℃.

   8.   Drizzle olive oil and place sliced truffles to finish.

¹⁹ SEAFOOD & TOMATO QUICHE

해산물 & 토마토 키슈

ingredients - ∅ 16cm 2ea

파트 아 브리제(57p) or 파이 반죽 – 글루텐 프리 (61p)

GLUTEN FREE   NUTS FREE

아파레유

생크림 125g
우유 63g
달걀 75g
소금 2g
케첩 30g
후추 적당량

가르니튀르

관자
새우
양파
마늘
방울토마토
브로콜리
청양고추

크림치즈
체다치즈
고다치즈
올리브오일
소금
후추
로즈마리

장식물

바질
드라이토마토

PATE A BRISÉE(57p) or PIE DOUGH - GLUTEN FREE(61p)

GLUTEN FREE   NUTS FREE

APPAREIL

125g Fresh cream
63g Milk
75g Eggs
2g Salt
30g Ketchup
QS Ground pepper

GARNITURE

Scallops
Shrimps
Onion
Garlic
Cherry tomatoes
Broccoli
Cheongyang pepper (or other spicy chili pepper)

Cream cheese
Cheddar cheese
Gouda cheese
Olive oil
Salt
Ground pepper
Rosemary

DECORATION

Basil
Dried tomatoes

APPAREIL

1-1

1-2

GARNITURE

2

3

4

5

FINISH

6

7

Process

아파레유	1.	볼에 모든 재료를 넣고 혼합한다.
가르니튀르	2.	마늘과 양파는 올리브오일을 두른 팬에, 관자는 버터를 두른 팬에 볶는다. 새우와 브로콜리는 깨끗하게 세척한 후 끓는 물에 데친다. 크림치즈는 사방 1cm 크기로 자른다. 청양고추는 잘게 썬다.
	3.	방울토마토를 반으로 가른 후 팬에 고르게 펼쳐준다.
	4.	올리브오일, 로즈마리, 소금, 후추를 적당량 뿌려준다.
	5.	골고루 버무린 다음 200℃ 오븐에서 10분간 굽다가 150℃로 낮춰 15분간 굽는다.
마무리	6.	80% 소성한 셸 안에 준비한 가르니튀르를 채운 후 체다치즈와 고다치즈를 적당량 올린다.
	7.	아파레유를 가득 붓고 175℃ 오븐에서 약 40분간 굽는다. 바질, 드라이토마토를 올려 마무리한다.

* 드라이토마토 - 방울토마토를 얇게 슬라이스한 후 30보메 시럽 (물 1000g, 설탕1350g)에 담궜다가 꺼내 50℃ 오븐에서 충분히 건조시켜 사용한다.

APPAREIL	1.	Mix all ingredients in a bowl.
GARNITURE	2.	Stir fry garlic and onion with olive oil. Stir fry scallops in butter. Thoroughly clean shrimps and broccoli and blanch in boiling water. Dice cream cheese in 1cm cubes. Finely chop Cheongyang pepper.
	3.	Halve the cherry tomatoes and spread out evenly on a baking tray.
	4.	Drizzle olive oil and sprinkle rosemary leaves, salt, and pepper.
	5.	Toss to coat evenly. Bake for 10 minutes at 200℃, then reduce heat to 150℃ and continue to bake for 15 minutes.
FINISH	6.	In 80% baked tarte shells, fill with the garnitures and top with a moderate amount of cheddar and gouda.
	7.	Top it off with the appareil, and bake for 40 minutes at 175℃. Finish with basil leaves and dried tomatoes.

* To make dried tomatoes, slice cherry tomatoes thinly and soak in 30 Baume syrup (water 1000g, sugar 1350g). Dry sufficiently in a 50℃ oven.

PATISSERIE
by
GARUHARU

TASTE

계절에 따른 가장 좋은 재료를 탐구합니다. 이렇게 찾은 재료의 맛이 디저트에 충분히 표현되었는지, 맛의 밸런스가 조화로운지 확인합니다.

We explore the best ingredient for each season. We make sure that the taste of the ingredients found is sufficiently expressed in the dessert and that the balance of the taste is harmonious.

TEXTURE

테크닉적으로 좋은 텍스처를 완성하는 것에 더해 하나의 디저트 안에서 단조롭지 않은 다양한
텍스처의 재미를 느낄 수 있도록 구성합니다.

In addition to mastering a technically good texture, we try to compose
to experience the fun in various textures in one dessert that will not be
monotonous.

DESIGN

디저트의 맛을 연상시킬 수 있는 포인트를 담은 간결한 디자인을 추구합니다.

We pursue a simple design with an emphasis that resembles the taste of the dessert.

Team GARUHARU

가루하루의 시작부터 지금까지 새로운 시도와 도전에 늘 열정적으로 동참
해주는 가루하루 팀에게 고마운 마음을 전합니다.
성실하고 열정적인 재능 있는 동료들과 함께여서 새로운 시도를 주저하지
않고 무모했던 도전의 과정을 즐길 수 있었습니다.

I would like to express my gratitude to Team
GARUHARU for their passionate participation in
new attempts and challenges since the beginning of
GARUHARU. Team GARUHARU for their passionate
participation in new attempts and challenges.
With these talented teammates who are sincere and
enthusiastic, I was able to enjoy the process of reckless
challenges without hesitating to try new things.

GARUHARU

SEOUL

GARUHARU MASTER BOOK SERIES 2

TARTE by GARUHARU

| First edition published | July 1, 2020 |
| Sixth edition published | October 10, 2023 |

Author	Yun Eunyoung
Assistant	Kim Eunsol
Translated by	Kim Eunice
Publisher	Han Joonhee
Published by	iCox Inc.

Plan & Edit	Bak Yunseon
Design	Kim Bora
Photographs	Park Sungyoung
Stylist	Lee Hwayoung
Sales/Marketing	Kim Namkwon, Cho Yonghoon, Moon Seongbin
Management support	Kim Hyoseon, Lee Jungmin

Address	122, Jomaru-ro 385beon-gil, Bucheon-si, Gyeonggi-do, Republic of Korea
Website	www.icoxpublish.com
Instagram	@thetable_book
E-mail	thetable_book@naver.com
Phone	82-32-674-5685
Registration date	July 9, 2015
Registration number	386-251002015000034
ISBN	979-11-6426-125-3